SEA CAVES of ANACAPA ISLAND

SEA CAVES of ANACAPA ISLAND

David Bunnell

McNally & Loftin, Publishers

James Whistler

Sea Caves of Anacapa Island

McNally & Loftin, Publishers
P.O. Box 1316
Santa Barbara, CA 93102

Printed by Kimberly Press, Inc.

All photographs were taken by the author

ISBN: 0-87461-093-1

NOTICE!

The maps in this book are not nautical charts.
They are not intended for navigation.

Library of Congress Cataloging-in-Publication Data

Bunnell, David Edward, 1952 -
Sea Caves of Anacapa Island/David Bunnell, -- 1st ed.
cm.
ISBN 0-87461-093-1 : $15.00
1. Marine caves -- California -- Anacapa Island.
I. Title
GB601.8.B85 1993 551.4'47'0979491 -- dc20 93-22206
CIP

CONTENTS

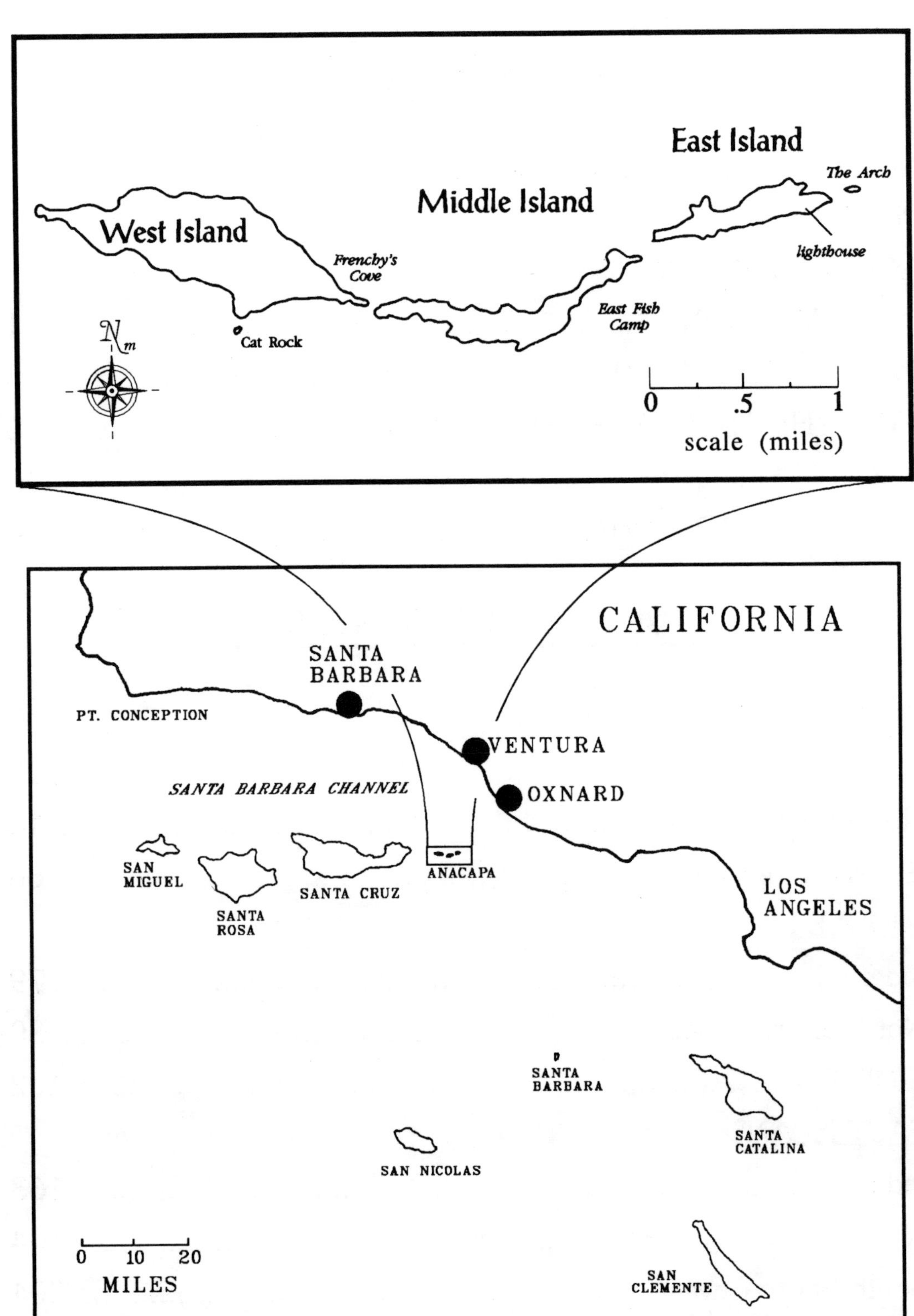

Portion of southern California coast showing Anacapa and the other 7 Channel Islands. Enlarged portion details the 3 islets which comprise the Anacapas.

PREFACE

I suspect that most of you who are examining this book already know where Anacapa Island is, but I've enclosed the map on the facing page to put things in perspective. Anacapa is the most visited of the islands which comprise Channel Islands National Park, and is certainly the closest and most accessible. It is a rugged and beautiful spot, and offers many recreational activities. One that is increasingly popular is exploring the marine caves that dot the shoreline.

Caves are probably not the first thing that come to mind when one thinks of Channel Island National Park. Yet even a brief excursion along the steep volcanic cliffs which comprise much of Anacapa Island will reveal dozens of mysterious cave entrances. How far in do these wave-cut caves extend? Is it true that some caves go all the way through the island? What's to be found in these caves? This book is meant to answer some of these questions and unravel some of the mystery. With the help of other members of the National Speleological Society (an organization of cave scientists and explorers), I've explored and mapped all the caves on the three islets that comprise Anacapa Island. For our purposes, a cave was defined as any opening extending 30' or more into the cliff that was enterable by a human being. By this definition, there are 135 caves on this small island! Like its neighbor Santa Cruz Island, Anacapa Island houses a world-class collection of sea caves. This manuscript is the first comprehensive survey of the island caves yet undertaken and should be of interest to scientists (*e.g.*, geologists, marine biologists), yachtsmen, sea kayakers, and divers.

DISCLAIMER: I AM NOT ADVOCATING THE EXPLORATION OF SEA CAVES TO ANYONE. THIS BOOK IS SIMPLY MEANT TO INFORM THE INTERESTED READER ABOUT THE CAVES AND POINT OUT THE DANGERS INHERENT IN EXPLORING THEM. SEA CAVES CAN BE DANGEROUS BOTH ABOVE AND BELOW WATER AND EXPLORATION IS DONE AT YOUR OWN RISK!

ACKNOWLEDGMENTS

This book is the result of over 3 year's field work by myself and other cavers, most of whom are members of the the National Speleological Society. I would like to thank the following individuals who participated in the survey of the caves on Anacapa Island: Djuna Bewley, Ann Bosted, Peter Bosted, Rich Breisch, Dan Clardy, Phil Darling, Don & Lisa DeLucia, Mike & Pam Doe, Lysa DeThomas, Ernie Garza, Bill Liebman, Don & Susan Morris, Rick Niemi, Matt Oliphant, Nancy Pistole, Carl Reuter, Bob Richards, Lori Schultz, and Carol Vesely.

Special thanks to Channel Islands National Park for providing support in the last months of the project.

Thanks are also due to two friends whose efforts helped the survey: Derek Hoyle, for donating his time and mechanical experience to work on the *Island Caver*'s towtruck; and Ed Schultz, who provided me and my fellow seacavers a place to stay in Santa Barbara during the years I had to commute to the islands from my home in Santa Cruz. Above all, this book is dedicated to my parents, whose generosity allowed the purchase of my research vessel, the 20' long *Island Caver*.

INTRODUCTION

The rugged Anacapas, edged by sheer basaltic cliffs and riddled with caves and arches, present some of the most spectacular scenery in the Channel Islands. With every scenic vista, the visitor is reminded that this is, truly, an island. Located some 15 miles south of Ventura, California, the Anacapas consist of 3 rocky islets strung out in a chain 5 miles long. Rocky cliffs compose the entire shore line, with a few small beaches appearing at low tides. In these cliffs are found an amazing diversity of wave-cut caves.

It was these caves, and the scarcity of knowledge about them, which first attracted me to these islands. In the previous six years I'd been exploring and mapping sea caves on Santa Cruz Island, turning up numerous large and interesting caves. Maps and descriptions of those caves appeared in *Sea Caves of Santa Cruz Island*, published in 1988. With Santa Cruz completed, I turned my efforts to the Anacapas. As was the case for Santa Cruz island, there was little published information on the sea caves of Anacapa. Lorenzo Yates visited some of the caves (apparently Frenchy's, Moss Cave, and Island View Cave among them, although he did not use these names) and wrote about them in the *American Geologist* in 1890. In 1952 Phil Orr of the Santa Barbara Museum of Natural History led a small expedition to explore and map caves in the Channel Islands. His unpublished manuscript describes 14 caves on Anacapa and suggests that there are "about 50" caves altogether. A glance at the current Park Service map of Anacapa notes only two caves, Frenchy's and Cathedral Cove - certainly two of the larger and more interesting caves. However, our survey indicates that there are actually 135 caves on the island, and many more if you count those less than our 30' criterion. Many of these are lengthy and surprisingly complex, including high-and-dry caves, "mazey" caves, two-level caves, caves with chambers opening only at low tide, and caves with long, sinuous, dark passages. Altogether, we surveyed 20, 590' of cave passage on the island, or almost 4 miles of humanly traversible cave passage! Although the "average" Anacapa cave is somewhat smaller than those on Santa Cruz (152' vs. 200'), the second (Catacombs, p.182) and third (Cathedral, p. 170) longest caves from the combined Channel Islands surveys are found on Anacapa Island. Our work now continues on Santa Rosa island, where many large caves have been found including three with "pits" or roof collapses into large chambers below. One such example of this unusual feature is found on East Anacapa (Half Gone Cave, p. 188).

History

The Anacapas were visited seasonally by the Chumash Indians, who fished its waters but apparently never settled permanently as they did on neighboring Santa Cruz Island. The name Anacapa is said to be derived from the Chumash word "Eneepah", meaning "ever-changing", and first appeared on charts made by the British navigator George Vancouver in the 1790s. Although the island has always been under government ownership, a series of leases were granted to various individuals for farming and grazing use. One of Anacapa's most notable residents was "Frenchy" LaDreau, who occupied some buildings (since removed) at present-day Frenchy's Cove from 1928-1956, and served as an unofficial National Park Service representative. He acquired water from seeps in the caves to the west, probably Frenchy's and Moss Caves. The island was designated part of Channel Islands National Monument in 1938, which then became Channel Islands National Park in 1980. The park also includes Santa Barbara, Santa Rosa, San Miguel, and the eastern portion of Santa Cruz Island.

There is little "history" to be told of the caves on Anacapa other than Frenchy's use of the caves to the west for water. The same two caves apparently showed evidence of use by Indians when visited by Lorenzo Yates on a voyage in 1890. He noted remains of a rope woven from "sea hair" and an artificial basin built to collect the water from the seeps. While there are numerous other "dry" caves on the island which might have provided shelter for visiting Indians, there is no evidence (e.g., midden sites) that they were used in this way.

As someone who loves the caves I would request one thing from all readers: to respect the caves. This I mean from both a safety standpoint and a conservation standpoint. **Exploring these caves can be quite dangerous**, as the ocean and its movements can be quite unpredictable. The same power that formed the caves could easily play havoc with an unwary explorer. Second, I ask that you respect the aesthetic qualities of the caves by trying to impact them as little as possible, which means not defacing the walls in any manner. In fact, it would be illegal to do so under that laws governing the Channel Islands National Park. A good guideline is found in the motto of the National Speleological Society, of which I am a member of long standing: "Take nothing but pictures, leave nothing but footprints".

The inner chamber of Frenchy's Cave

EXPLORING SEA CAVES

Included with each cave description is data pertinent to the conditions which may be expected when exploring that cave: e.g.., is the cave suitable for exploration by dinghy?, are lights needed?. In our explorations, the caves were typically reached and explored by snorkeling and/or "tubing", i.e., using an inner tube for flotation. We were also equipped with a full wetsuit, fins, strong waterproof (dive) lights, and a helmet. I can't overemphasize the value of a helmet. Not only might you bump your head in a low section of passage, but a swell entering the cave might propel you into the rock with great force. Some of us have taken pretty good whacks during exploration and have been extremely grateful for the protection our helmets afforded.

Sea caves are explored in one of 3 ways once you reach the entrance: you walk or wade, you swim or snorkel, or you use a small watercraft such as a dinghy or kayak. I've not yet found any caves of significant length on Anacapa that are wholly submerged and hence, require scuba gear. If you find any, I'd sure like to know about them.

Some caves you can walk through at any tide level. These are "relict" caves, those which are high enough in the cliff that they are no longer actively forming. The floors of these caves are always dry. Anacapa has 12 such caves, 8 of which also have an actively forming lower level. Many of the island's caves empty out sufficiently at low tide that you can walk or wade shallow water to explore some or all of the area beyond the dripline. On the other hand, many caves are floored by swimming-depth water even at low tide. Many of the latter caves have passages and ceiling heights large enough to permit exploration by dinghy or kayak. In some cases these caves will have a beach in the rear where one could land a kayak or dinghy if seas are calm (if they aren't, what are you doing in there anyway!?). I wouldn't recommend a motorized craft in any of these caves. Not only is it very damaging to the cave's ecosystem to fill it with exhaust fumes, but most caves have submerged rocks that could play havoc with props.

Two factors which have the greatest bearing on sea cave exploration on any particular occasion are the swell level and the tide level. Under no circumstances should one enter a sea cave when a large swell is rolling in! In January, 1992, two people were killed when they entered Cathedral Cave in high-swell (about 8' !) conditions. Search parties combing the area said the cave would appear fairly calm until a huge swell came in and filled the cave to its ceiling! Most caves have an amplifying effect on prevailing swell levels, since the passages tend to go from wide to narrow. . Since swell directions are fairly predictable, one may simply choose to visit caves which are protected from the prevailing swell. The west-northwest swell typical around Anacapa Island precludes entry into many caves when attaining heights of 4' or more, particularly along West Anacapa. However, even a 2' swell sometimes produces problems if it aims directly into a cave. Southwest swell patterns are also common, and favor caving on the north shore. The calmest seas are generally observed during June through October, and hence, these seem to be the best months for exploring sea caves on the islands. The period January through May often brings swells of 3 to 6 feet, meaning only the most sheltered of caves will be safe to explore. The seas, and hence, the caves, tend to be calmest in early morning before the winds pick up.

Variations in tide level may change the water level in caves by as much as 7'. Generally speaking, the lower the tide, the higher the cave's ceiling if you are floating in! Moreover, in many cases a cave may only be seen in its entirety at low tide (unless you're on scuba) as passages with low ceilings open up. An additional benefit of low tide exploration is that more marine life will be exposed in the caves. Thus, the wise sea caver never leaves home without his or her tide tables. The necessity of low tide exploration for any particular cave is addressed in its description.

ACCESS POLICIES

All the caves fall under National Park Service jurisdiction. Presently, there are no rules against visiting any of the caves, with the exception of the caves in the Pelican Closure Area on West Anacapa. This area is a major pelican rookery, and access to Island View, Confusion, Pinnacle, Moss, and Frenchy's Caves is currently restricted to the months of November and December. However, check current regulations showing the boundaries and time periods for the closure.

Additionally, please note that, while it is permissible to land along the shorelines of Middle and West Anacapa, no one is permitted to go up on the island (i.e., on the cliff tops) without a permit.

Caves should not be visited when seals are hauled out on beaches inside. The Marine Mammal Protection Act of 1972 stipulates that these animals may not be harassed, hunted, captured, or killed. According to the act, harassment may be interpreted as any activity that alters the behavior of the animals. Seals are commonly found in only a few caves, most notably Seal's Refuge in Cathedral Cove and in Sea Lion Cave on West Anacapa. Additionally, caves should be avoided April through July if cormorants are nesting on ledges in the entrance zone. The approach of a kayaker may frighten the birds, who may sometimes knock their nests down as they flee. While there are currently no specific regulations to protect these birds, please give them the space they need. If you would like further information, please contact the staff at the Channel Islands National Park visitor's center.

KAYAKING

Sea kayaking is an outstanding way to explore the coastal cliffs with their numerous arches and caves. We made good use of our Ocean Kayak Scuppers, which were invaluable for scouting. They had the advantage of being "no roll" kayaks (i.e., your body isn't inside the boat), making them MUCH safer were one to be flipped inside a cave. Polyethylene construction is also an advantage for the potentially rough treatment a sea cave might deliver. The best caves for kayaking are either those that allow a "through-trip" between entrances, or those which are so spacious that one can turn a 12-18' boat around inside. Half-Gone and Cathderal Cave on East Island are a must for kayakers. On Middle Island, kayakers may wish to check out Complex Chasm and Treasure Chest Cave, while on West Island, Impalement, Birdshower, Three Bedroom & a Bath, Three Fingers, Green Shelf, Island View, Confusion, and Frenchy's Caves are of special interest. At low tide, many caves have small beaches or rock benches nearby where one might land and explore caves on foot, or if conditions are sufficiently calm, one may land inside the caves themselves. Note also that many caves are best visited by kayak at moderate or high tides, since lower tides expose rocks or even leave the cave high and dry. At any tide level, though, there is much for the kayaker to explore on Anacapa.

DIVING THE CAVES

Many of the caves on the island are of interest to divers, as it is no secret that some may house populations of lobster. Currently, the north shores of East and Middle Anacapa are preserves, so no lobster may be taken from these caves at any time. Some of the caves in these areas, such as LaGrieta, afford a good opportunity to see undisturbed lobster populations. I have not indicated any lobster "hot spots" in my descriptions, but the maps will give an idea of which caves might contain water depths suitable for diving. Most diving in these caves involve what is basically an open-water night dive as long as some air-filled passage remains overhead, with the added danger of potentially surgey conditions in narrow confines. ALL THE CAUTIONS ABOUT SWELL CONDITIONS ABOVE SHOULD BE PARTICULARLY HEEDED BY DIVERS!. Some caves have passage continuations or side passages which are entirely submerged (possibly varying with tide levels) and thus fall within the realm of cavern or cave diving. No one should attempt such diving without specific training in cave diving! A properly trained cave diver carries at least three sources of light, lays a continuous line to the entrance, and begins exiting the cave after having consumed no more than a third of his air supply. Those interested in more information about training in cave diving techniques should contact the NSS Cave Diving Section, PO Box 950, Branford, FL 32008-0950.

USING THE CAVE DESCRIPTIONS AS A GUIDE TO EXPLORATION

For each of the 135 caves surveyed on the island, there is a separate description, which includes its approximate location, its length, the number of entrances (which may vary with the tide level), the conditions involved in exploration, and a basic description. For each island, the caves have been numbered sequentially from west to east on the north shore and continuing from east to west on the south shore. These numbers correspond to those on the location maps at the beginning of each section. The photographs should aid in locating the caves visually, since many are clustered in close proximity.

One of the most important considerations when using the descriptions is to note the tide level the map is based on. Conditions may be very different at high tide - in fact, some of the caves one walks through at low tide may require scuba at high tide! I have tried to address the relative importance of the tide in the "conditions" section. This section also indicates whether a light is necessary to navigate in the cave. Very few caves actually require lights but they are useful in many cases for examining tidepool life and to avoid urchins, which floor many of the caves. Finally, hazards specific to each cave are noted, such as submerged rocks, or a tendency to amplify prevailing swell.

HOW THE CAVES FORMED

Formation of the island

Anacapa's story began 20-25 million years ago in the Miocene epoch. Sea floor spreading under the Pacific Ocean was pushing the Farallon plate northeast, where it was subducted under the North American plate. This caused large quantities of lava to be extruded near the plate margin. Over a period of about 15 million years successive layers over 8000' thick accumulated.

The next chapter of the story began with a period of extreme regional uplift, around 2-5 million years ago. Uplift occurred along fault zones, resulting in a block containing the present day northern Channel Islands (San Miguel, Santa Rosa, Santa Cruz, and Anacapa) being thrust upward. Geologists describe the islands as a block-faulted east-west trending anticline (i.e., an upwardly convex fold), with both flanks exposed on Santa Cruz and Santa Rosa Islands and only the northern flank exposed on Anacapa. This anticline is aligned with and structurally related to the Santa Monica Mountains. In fact, the volcanic rocks which make up Anacapa are called the Conejo Formation, named for a type locality in the Santa Monicas. Geologists describe this formation as "black to dark-red vesicular and porphyritic basic extrusive and pyroclastic rocks, consisting of massive-to-thinly bedded lavas, autobrecciated flows, lenses of lapilli tuffs and tuff breccias, and beds of volcanic breccias and agglomerates" (Scholl, 1959). All the caves on Anacapa are formed in this unit. The diversity of volcanic forms found in this unit is reflected in the caves.

Formation of the caves

The uplift regime produced sea cliffs on Anacapa ranging from 250 to 325' high. During this process, severe stresses were placed on the rock, forming hundreds of localized faults and fractures. Most of these are high-angle strike-slip faults oriented perpendicular to the long axis of the island (Scholl, 1959). High angle faults are those which show significant dip rather than running vertically. Many of the faults forming the caves dip at angles of 20° or more. Strike-slip faults are those where the displacement is lateral, along the strike of the fault, rather than up or down, such as in the dip-slip fault that caused the uplift of the islands.

The faults created zones of weakness in the cliff-faces. As the surf pounds into these areas, loose rock is removed. As the sea reaches into the fissures thus formed, they begin to widen and deepen due to the tremendous force exerted upon a confined space. Blowholes (partially submerged caves which eject large sprays of seawater), attest to this process. One of these appears at Landing Cove at certain tide levels. Scholl (1959) reports standing atop the cave-riddled eastern tip of West Island at high tide and feeling the impact of the surf entering the caves through nearly 100' of overlying rock! Even on a fairly calm day, one can hear a resounding thud from inside the caves at high tide. Adding to the erosive power of the waves is suspended sand and rock. Most of the cave walls are irregular and chunky, reflecting an erosional process where the rock is fractured bit by bit. However, some caves have portions where the walls are smoothed and rounded. These are typically found in areas floored by cobbles, and probably results from the swirling about of the cobbles in heavy surf. The effect would be much like that which forms potholes, but in the caves the scouring action is more horizontal than vertical.

Often caves will have branching passages where one fault trend intersects another. A remarkable example of this is Catacombs Cave on East Anacapa. Here, 7 different faults or fractures have intersected to allow the formation of a cave with a maze-like pattern. Caves with more than a single entrance usually have passage developed on more than one fault, with the exception of tunnel-like caves like the Cathedral Arch.

Cave formation has reflected an interplay of ongoing uplift and changes in sea level. The elevation of the island relative to sea level has been steadily increasing. This accounts for the tall cross sections typical of most of the caves. Many of the caves contain underwater canyons 20-30' deep, and probably represent erosion occurring at lower sea level stands, 20,000 years ago or more. By contrast, the dry upper levels seen in some caves probably do not reflect higher sea levels but rather, a period of relative stability in the uplift regime. The marine terraces seen atop East and Middle Islands reflect the same process. Sea levels have been fairly stable over the last 6,000 years, so much of the cave passage on the current "active" level probably formed in that time.

Rain water falling on the island's surface may also influence cave formation. Carbonic and organic acids leached from the soil may aid groundwater in weakening the lower volcanic units. This may also result in the dissolution of carbonates and sulfates and their deposition within the caves. These appear on the cave walls and ceilings as small stalactites or flowstone (carbonates) and white crusts (gypsum). Gypsum is seen in several of the caves with permanently dry regions, such as Teardrop and Frenchy's. In the latter cave, daylight has promoted growth of a green algae in the gypsum matrix. Stalactites were seen in Respiring Crevice and orange flowstone may be seen in Little Sunbeam Cave.

Sea caves play an active role in shaping the Anacapas. They undercut the cliffs and promote seacliff retreat. The passages between the islands probably resulted from the collapse of sea caves which had cut through the narrower portions. In his 1890 manuscript Yates depicted an arch connecting the east tip of West Island (containing Long Beach and Shortcut Caves) with Middle Island. The arch has since collapsed, with its remains blocking what will become a third passage between islands.

Further reading:

Moore, D.G. Origin and development of sea caves. *National Speleological Society Bulletin,* 16: 71-76, 1954.

Scholl, D. W. Geology and surrounding recent marine sediments of Anacapa Island. Ph.D. dissertation, University of California, Santa Barbara, 1959.

Yates, G.Y. Notes on the geology and scenery of the islands forming the southerly line of the Santa Barbara Channel. *The American Geologist*, January 1890.

ABOUT THE MAPS

The maps in this book are based on actual compass and survey tape measurements taken along with a sketch of the cave made *in situ*. This involved snorkeling or "tubing" into or to the cave, establishing (but not permanently marking) specific points on the walls, and determining the distance and compass bearing between these points. A plan view of the cave was sketched at the same time and the points noted on the sketch. This process was rendered more difficult in the wet caves, owing to the presence of swell and the resulting difficulty in keeping one's position on a survey station. Divers' underwater slates were used for making the sketches in the wet caves. Dive compasses were generally used for bearings, and though somewhat less precise than the optical sighting compasses normally used in cave surveying, were the only instruments which could survive the conditions. Nevertheless, in those rare caves where a survey loop was made, the loop closures were surprisingly accurate (1.0%).

All but 5 of the maps were produced by the author using Corel Draw! in conjunction with a flatbed scanner and a laser printer. The maps of Cathedral, Catacombs, Purple Palace, and Sea Lion were inked by Bob Richards and the map of Half-Gone Cave was inked by Carol Vesely. Their maps use the same symbols as mine, except that their cliff lines are represented by lines of small "xxxx" instead of thick gray lines.

READING THE MAPS

All the maps follow the typical standards for cave mapping, depicting a plan view, or what one would see if looking into the cave from above. Note that in doing this, the walls are drawn at their widest. In some caves the walls are undercut and the passage may seem much narrower, but these instances are shown as changes in ceiling height on the maps. Most of the maps also show cross-sections, graphically indicating the passage shape and size in a plane perpendicular to the plan view. To give a better idea of the ceiling heights, I've added either a 6' tall sea cave explorer (in dry or shallow water) or a 3' high kayaker (in deeper water). Ceiling heights shown are based on estimates. Passage widths in the larger caves were usually measured to ensure the accuracy of the sketch. Finally, where caves are in close proximity, a survey was conducted along the cliff-face between them and they appear on the same map together in their proper orientation. In some instances no cliff survey was conducted due to rough condition and the shape of the cliffs and distances between the caves was estimated; if so, this is indicated on the map.

Certain types of information are found on each map and bear some further explanation:

Cave Name

With the exception of Frenchy's and Cathedral Caves, none of the remaining caves seem to have had any historical or generally accepted names. Rather than just referring to each as "Cave #x" in the maps and descriptions, I have assigned names which either a.) relate the cave to a nearby named feature, such as a point or anchorage (few of which exist on Anacapa), or b) describe a salient or distinctive feature of the cave, or c.) indicate something about the conditions or ambience during the survey. In choosing names we have avoided assigning personal names of the explorers (e.g., "Dave's Cave"), except in the case of Phil's Folly Cave, named after a sea caver who would fearless go into any cave regardless of conditions. Fortunately he survived all of them.

Scale

The scale may vary somewhat from map to map, and was chosen to be appropriate to the level of detail deemed necessary. Since some maps are reduced much more than others, it is important to examine the scale. The scale for all the caves is given in feet, as are all the ceiling heights, drop heights, and water depths.

North Arrow

On most maps magnetic north is towards the top, or if not, slightly rotated. While this means the orientation of the maps for north shore cave may be backwards to what one observes in approaching them, it is more consistent with map-making standards. Note that magnetic north in this region is about 15° to the east of true north.

Tide Level

The tide level given is that during which the survey of that cave was conducted. This number reflects an average as the actual level would have varied over the time (usually 20 minutes to over an hour) it took to survey the cave. It is important to remember that all of the information depicted on the map is relative to that tide level. Ceiling heights and particularly, the location of surf zones, will vary with the tides. Portions of some caves may be submerged at higher tide levels. I have tried to address this topic in the written description accompanying each map.

Length

The length is the total of the main passage plus any side passages. This is surveyed length and may be somewhat longer than you might compute by laying a ruler down over the cave on the map. This is because measurements may cross from one side of a passage to another and this becomes part of the length. In most instances the determination of length is straightforward but in some cases, such as a large chamber with radiating branches, its computation becomes a little hazy. Length may best be viewed as a relative guide to the size of the caves.

Symbols

Various features within the caves are depicted using standard cave mapping symbols, with some additional ones created from features unique to sea caves. The table on page 11 delineates most of the symbols found on the maps.

NOTES ON MAP SYMBOLS

Angular Rock Fragments: rockfall; when chunks are larger than a foot or so, they are designated with a breakdown symbol.

Bedrock: Area where the cave is floored with the host rock.

Breakdown: Rocks which have broken and fallen off the ceiling of the cave or the cliff above; typically found near the entrance zone.

Ceiling height: Estimated distance from floor or water's surface to the ceiling.

Change in Ceiling Height: Indicates abrupt variations in the height of the cave.

Cliff Face: Indicates portions of rock wall on the map which are outside of the cave.

Cobbles: These may vary in size within caves although not reflected in the symbol.

Dripline: Area at the cave entrance where water would fall from above if raining; used as the starting point for determining a cave's length.

Drop-Off: Typically these are from the top of a surge channel or upper level passage.

Drop-off height: Distance from a ledge to the floor or water's surface.

Pillar: A solid rock buttress which may be traversed on all sides.

Sea Caver: This 6' tall explorer is used to give scale in cross-sections with dry floor or shallow water. He\she wears a helmet and is usually shown carrying an inner tube.

Sea kayaker: This 3' high kayaker is used to give an idea of scale in cross-section with water-filled passage. Use of this symbol does not mean that the cave is optimal for kayaking - check the description.

Seawater: Water contiguous with the main ocean body

Slope: Lines fan out towards bottom of a slope, usually composed of cobbles or sand.

Surf Zone: Interface between ocean and dry land; varies with tide level.

Tidepool: Used to denote a tidepool or other standing water beyond the surf zone.

Underlying passage: A passage which extends underneath another cave passage.

Urchins: Indicates the presence of numerous urchins in shallow-water caves; be careful!

Water Depth: Included only where a reasonable estimate could be made, such as in shallow water, or when the bottom was seen through snorkel gear.

Water Depth Undetermined: Used in cross sections to indicate that no data on water depth was noted. When a reasonable estimate was made, it is included in the cross section by drawing a floor and a water line.

CAVE MAP SYMBOLS

Sea kayaker (3' tall)

Seawater, behind surf zone

Urchins

Ceiling height above water (feet)

Water depth (feet)

Ceiling height above water / depth below water

Drop off height (feet)

Sand

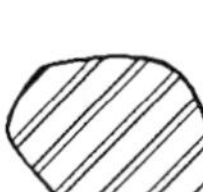
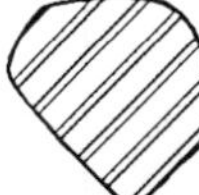
Tidepool or other still water

Pillar

Low side
High side
Drop-off

Edge of surf zone

Cobbles

Bedrock floor or shelf

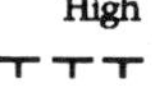

Abrupt change in ceiling height

Slope

Breakdown

Small angular rock fragments

Dripline (entrance)

Unsurveyed passage

Underlying passage

Cliff face (outside of cave)

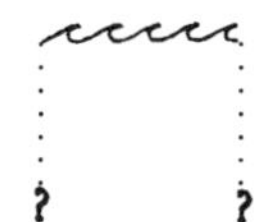
Water depth undetermined in cross-section

Magnetic north

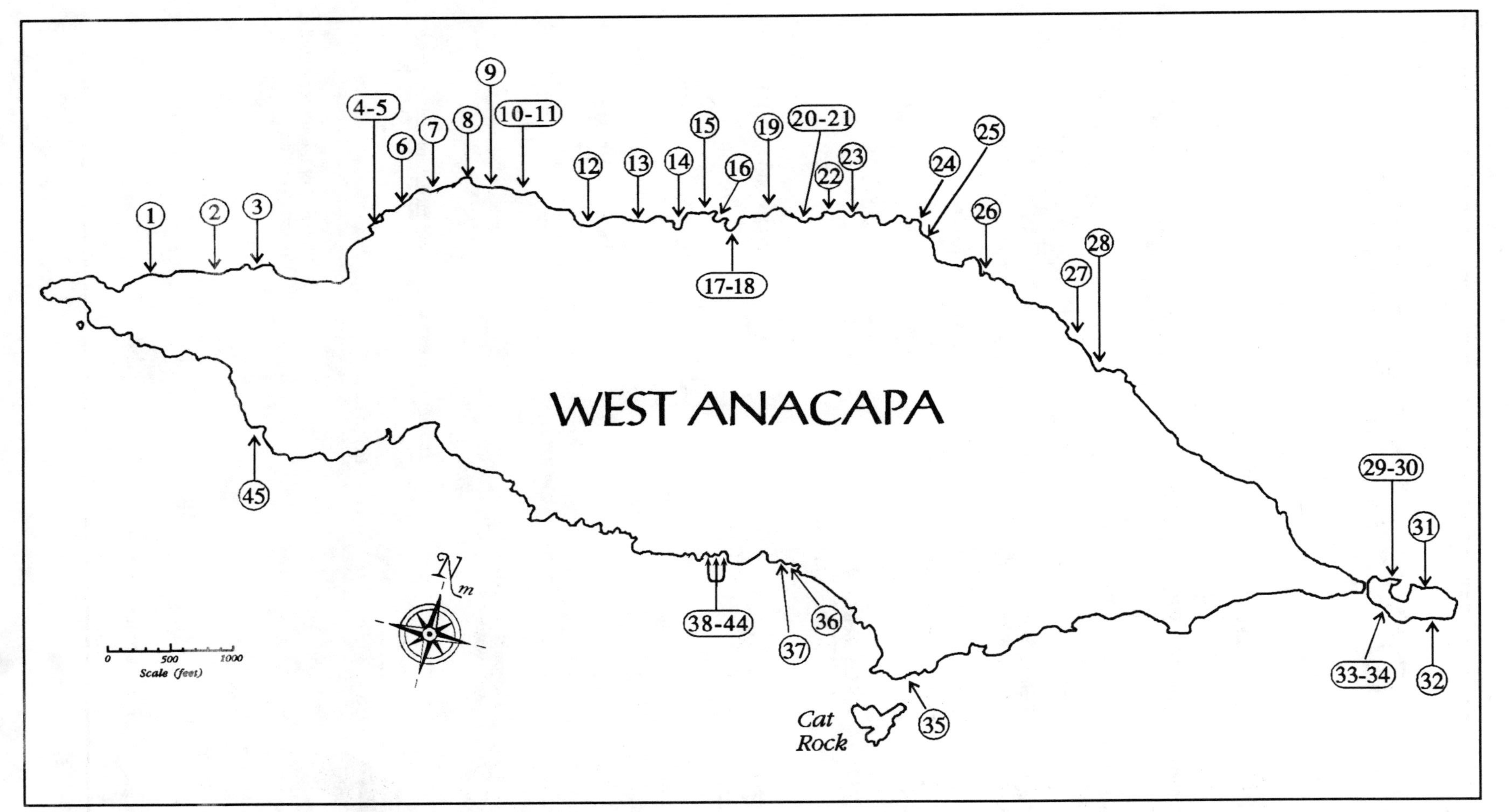
WEST ANACAPA
1
2
3
4-5
6
7
8
9
10-11
12
13
14
15
16
17-18
19
20-21
22
23
24
25
26
27
28
29-30
31
32
33-34
35
36
37
38-44
45
Cat Rock
N
m
0
500
1000
Scale (feet)

WEST ISLAND CAVES

NORTH SHORE

1 IMPALEMENT CAVE (230')
2 CRACK OF DOOM CAVE (155')
3 BIRDSHOWER CAVE (170')
4 WEST END FISSURE (58')
5 WEST END ARCH (50')
6 TWO BEDROOM & A BATH CAVE (291')
7 PURPLE PALACE CAVE (85')
8 TEARDROP CAVE (258')
9 THREE FINGERS CAVE (490')
10 IF YOU DARE CAVE (104')
11 TRUTH OR DARE CAVE (170')
12 ONE-SHOT CAVE (66')
13 SLAM DUNK CAVE (170')
14 NESTING CORMORANT CAVE (259')
15 NO FRILLS CAVE (41.4')
16 SEA LION CAVE (246')
17 CLUB-FOOT CAVE (36')
18 ROUGH GOING CAVE (70'+)
19 GREEN SHELF CAVE (158')
20 SUCKING SLOT CAVE (131'+)
21 TURBULENCE CAVE (50')
22 TRANQUILITY CAVE (114')
23 WOODEN LETTUCE CAVE (63')
24 ISLAND VIEW (272')
25 CONFUSION CAVE (242')
25a {large alcove & blowhole}
26 PINNACLE CAVE (164')
27 MOSS CAVE (100')
28 FRENCHY'S CAVE (421')
29 FRENCHY'S COVE CAVE (62')
30 FRENCHY'S TWIN ARCH (77')
30a {cave feature, 12' wide, 20" deep}
31 LEAPYEAR CAVE (85+')

SOUTH SHORE

32 LONG BEACH CAVE (372')
33 SHORTCUT CAVE (164')
34 SHORTY CAVE (43')
35 CAT'S SHADOW CAVE (86')
36 LITTLE SUNBEAM CAVE (90')
37 PHIL'S FOLLY CAVE (110')
38 CAT'S EYE CAVE #1 (58')
39 CAT'S EYE CAVE #2 (65')
40 CAT'S EYE CAVE #3 (40')
41 CAT'S EYE CAVE #4 (34')
42 CAT'S EYE CAVE #5 (84')
43 CAT'S EYE CAVE #6 (62')
44 CAT'S EYE CAVE #7 (58'
45 LONELY AT THE TOP (116')

Note: features listed in brackets { } are prominent cave entrances that open into caves smaller than our 30' survey criterion.

IMPALEMENT CAVE - 1

Location: 600' east of the west tip of West Island

Entrances: 1

Length: 230'

Conditions: This cave is best explored only in very calm seas. It gets dicey in a west or northwest swell. It is large enough to bring a kayak into, up to the first outcrop 150' or so in, but there is no place to land. A higher tide may actually make for easier exploration The rear portions are explorable by swimming only, and are hazardous due to rocks, surf, and thick kelp.

Description: The cave has a high, impressive entrance, 90' high and 32' wide. It is formed along a fault which dips eastward, evident in the entrance cross-section.The cave is a single water-filled passage which extends 230' to a wall, the ceiling height dropping gradually to a low of 20'. The water in the cave is deep, but we did not check the possibility of an underwater passage continuation.

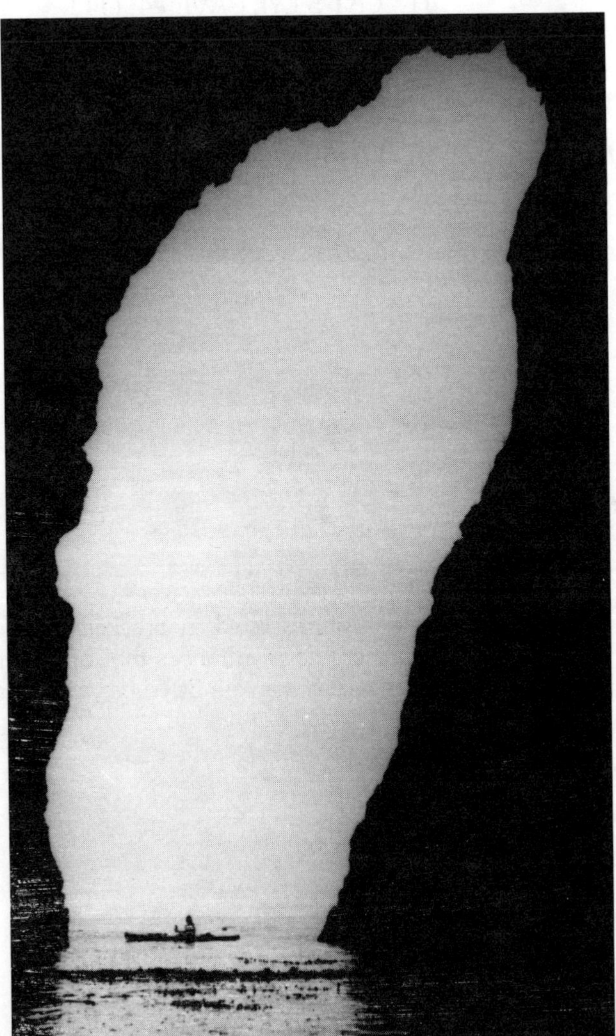

Looking out from Impalement Cave

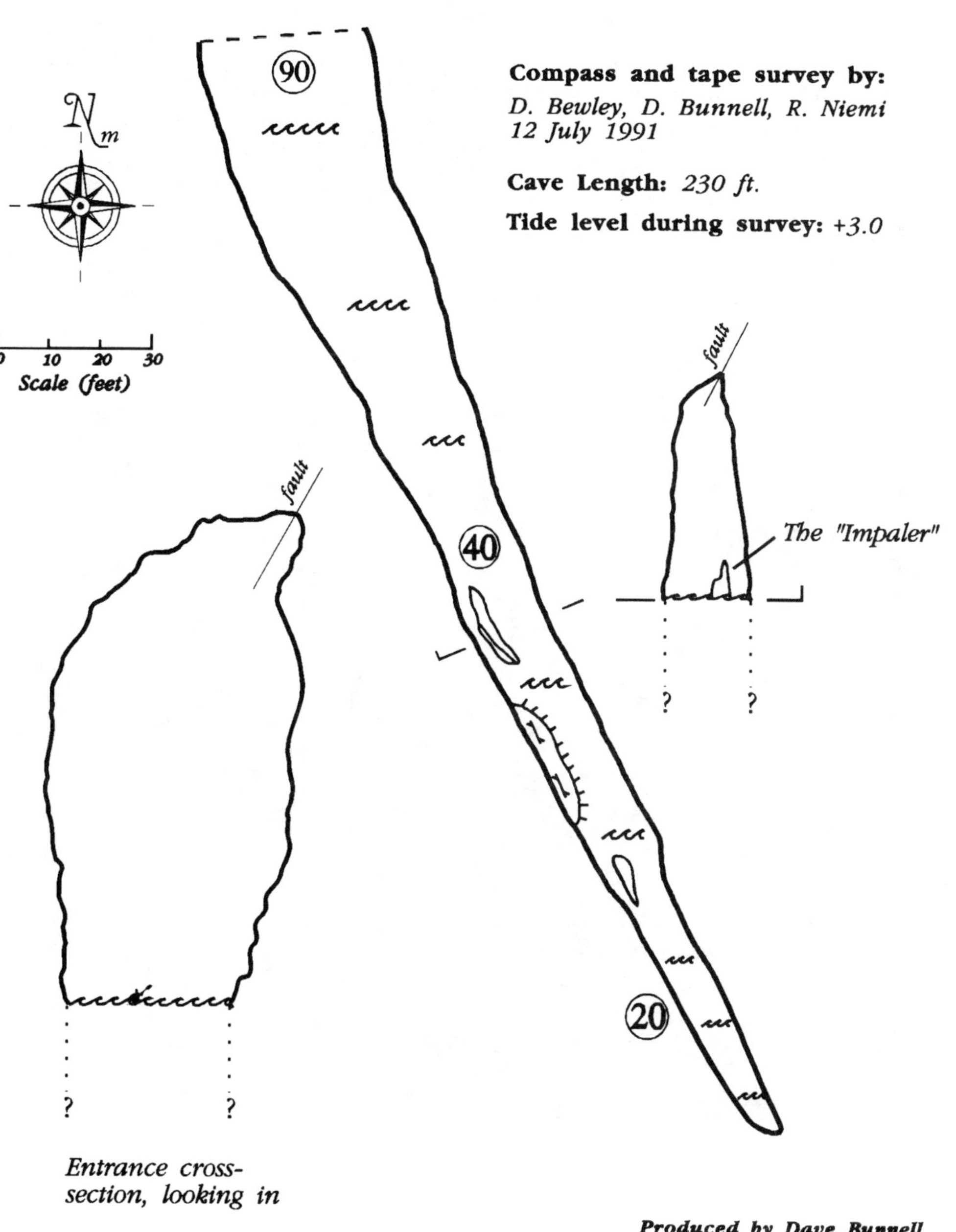
IMPALEMENT CAVE
West Anacapa Island, California
Channel Islands National Park
Compass and tape survey by:
D. Bewley, D. Bunnell, R. Niemi
12 July 1991
Cave Length: 230 ft.
Tide level during survey: +3.0
N
m
0 10 20 30
Scale (feet)
90
40
20
fault
fault
The "Impaler"
?
?
?
?
Entrance cross-section, looking in
Produced by Dave Bunnell

CRACK OF DOOM CAVE - 2

Location: About 1800' east of the west tip of West Island

Entrances: 1

Length: 155'

Conditions: The cave is water-floored throughout and can be entered a short way in a dinghy; exploring further requires swimming, a light, and very calm seas. Open to west-northwest swell. The cave's configuration seems to amplify the swell.

Description: The entrance is about 12' high x 20' wide and angles up to the right along the forming fault. After 105' the passage narrows to about a foot and a half wide, an area which was too intimidating to enter on our first visit. We returned on a very calm day to traverse another 50' of 6' wide passage beyond the narrows. Even on a calm day, we were pushed back and forth by a strong surge in this area. The water depth at the cave's entrance is 15-20' and this depth seems to be maintained for much of its length. The cave may well continue further underwater, requiring specialized cave diving techniques to explore. Ripples in the sand floor suggest the force of the surge in this cave.

Kayaking out of Crack of Doom Cave

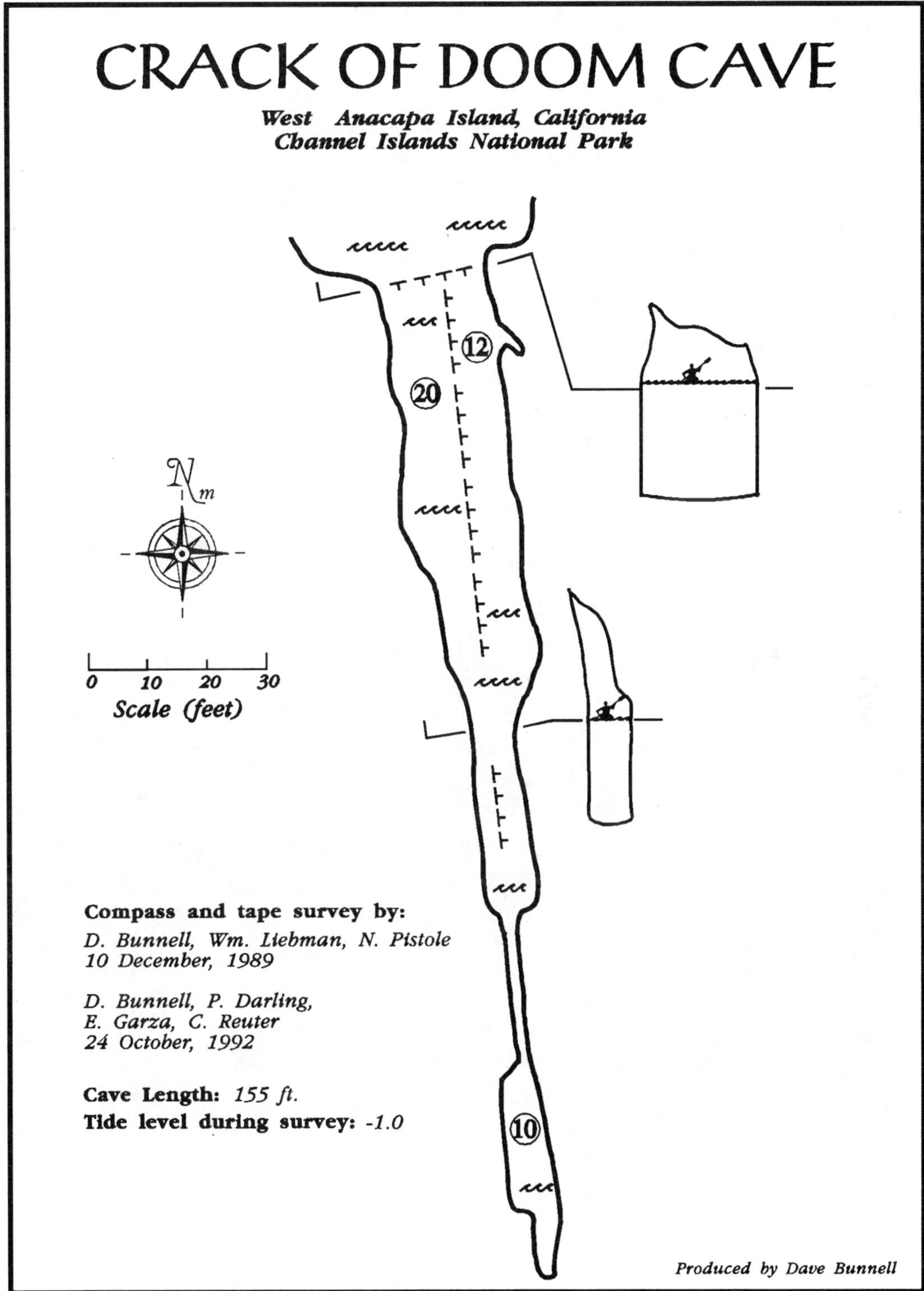
CRACK OF DOOM CAVE
West Anacapa Island, California
Channel Islands National Park
12
20
10
N m
0 10 20 30
Scale (feet)
Compass and tape survey by:
D. Bunnell, Wm. Liebman, N. Pistole
10 December, 1989
D. Bunnell, P. Darling,
E. Garza, C. Reuter
24 October, 1992
Cave Length: 155 ft.
Tide level during survey: -1.0
Produced by Dave Bunnell

BIRDSHOWER CAVE - 3

Location: About 1800' east of the west tip of West Island

Entrances: 1

Length: 170.3'

Conditions: The outer portions of the cave are readily explored in a dinghy. Exploration along the main passage requires swimming, and a light and low tide are needed to reach the rear portions.

Description: This interesting cave was named for the numerous birds that live on ledges above the entrance and their tendency to excrete on visitors passing underneath. The entrance is broad and arching, rising to perhaps 40'. Part way in, a finger of rock rises some 15' above the water on the righthand side. It is possible to scale this rock and get a nice view of the cave. Past the pinnacle, the cave narrows to 10' wide and extends some 80', climbing a couple ledges before ending in a beach with large cobbles.

Looking in the large entrance of Birdshower Cave

BIRDSHOWER CAVE

West Anacapa Island, California
Channel Islands National Park

Compass and tape survey by:
D. Bunnell, Wm. Liebman, N. Pistole
10 December, 1989

Cave Length: *170 ft.*
Tide level during survey: *-1.0*

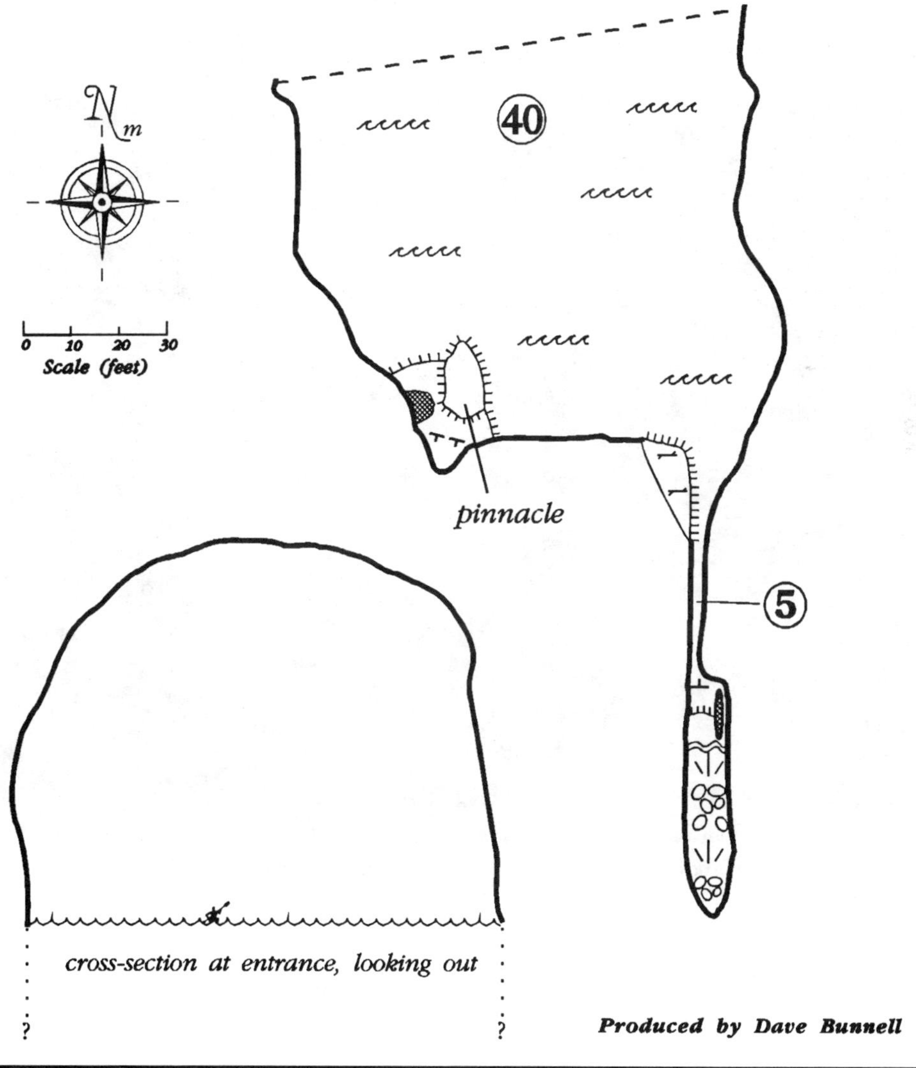

WEST END ARCH & FISSURE - 4,5

Location: 2200' from the west tip of West Island

Entrances: 2 (arch); 1 (fissure)

Length: Arch (50'); fissure (58')

Conditions: A trip through the arch may be made by dinghy or kayak at any tide level. The east exit can be tricky in a surge, though. The fissure south of it should be explored at low tide and with a light.

Description: The arch is a tunnel 50' long and generally about 20' wide and 12' high. The nearby fissure passage is about 15' high and 3 to 4' wide for most of its length, and is mostly dry at low tide.

Kayaking through the arch. Fissure cave on the right.

WEST END ARCH AND FISSURE CAVE

West Anacapa Island, California
Channel Islands National Park

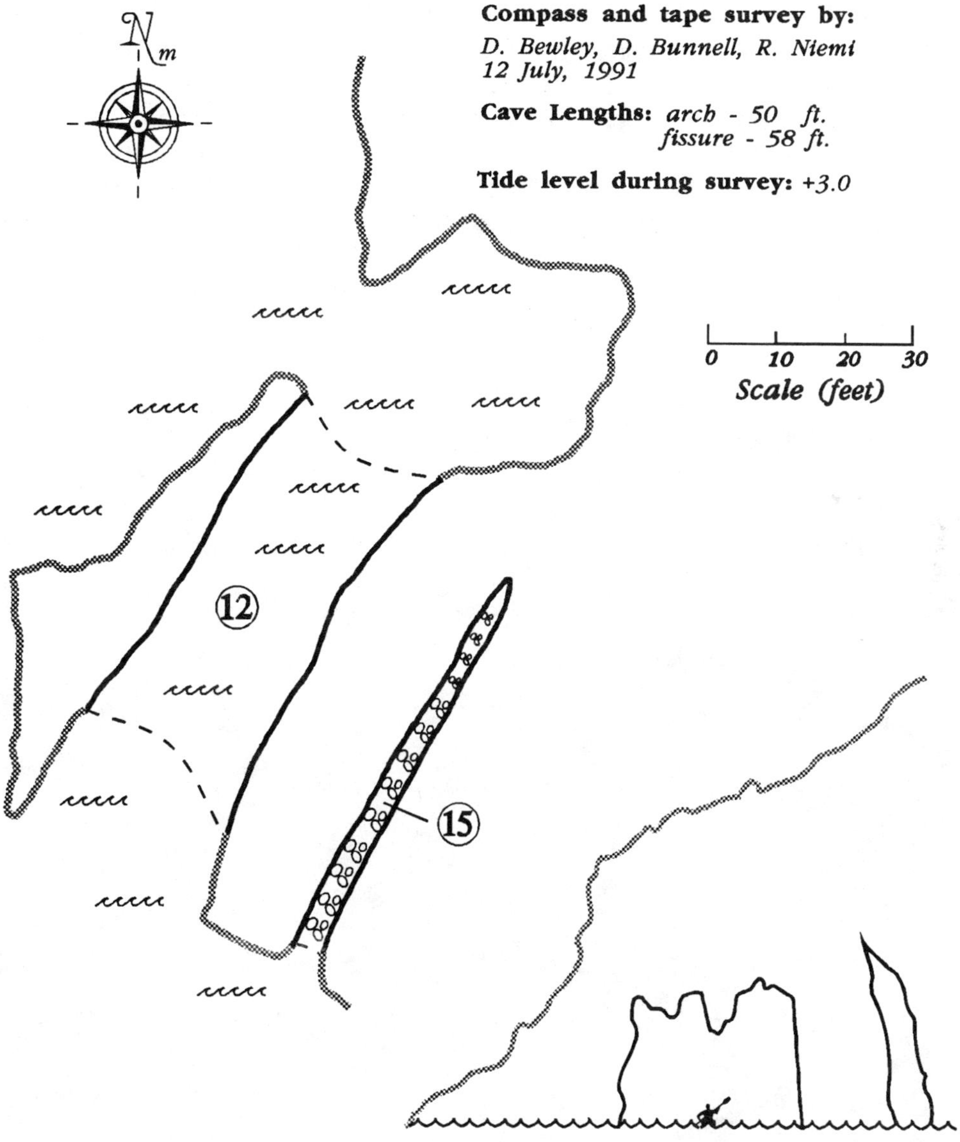

Compass and tape survey by:
D. Bewley, D. Bunnell, R. Niemi
12 July, 1991

Cave Lengths: *arch - 50 ft.*
fissure - 58 ft.

Tide level during survey: *+3.0*

N m

0 10 20 30
Scale (feet)

12

15

cave entrances as seen from the southwest

Produced by Dave Bunnell

TWO BEDROOM & A BATH CAVE - 6

Location: 2400' from the west tip of West Island

Entrances: 2

Length: 291'

Conditions: This cave is entirely water-filled and must be explored by swimming, kayak, or dinghy. A nice through-trip is possible, but at low tide, rocks exposed near the "bath" entrance make an exit here more problematic. The dark side passage requires a low tide, calm seas, and a light to explore; it is not large enough for dinghies.

Description: The larger (westerly) entrance lies behind a small baylet formed by an arm of rock which partly protects the cave from swells. This entrance is 50' wide and 20' high and leads into a 12' high chamber which intersects a second passage in the rear, along a fault trending 40-220°. To the left, this passage extends 140' to a second, smaller entrance. Passage heights vary from 12 to 20'. To the right from the entrance chamber, a passage extends along the same fault into total darkness. It is narrower, averaging some 6' wide and high and is walled with white sponge. During our survey, a kayak left afloat in the baylet outside the main entrance was pulled by surge through the cave and out the second entrance before it was recovered!

Kayaking into the "bath" entrance of the cave

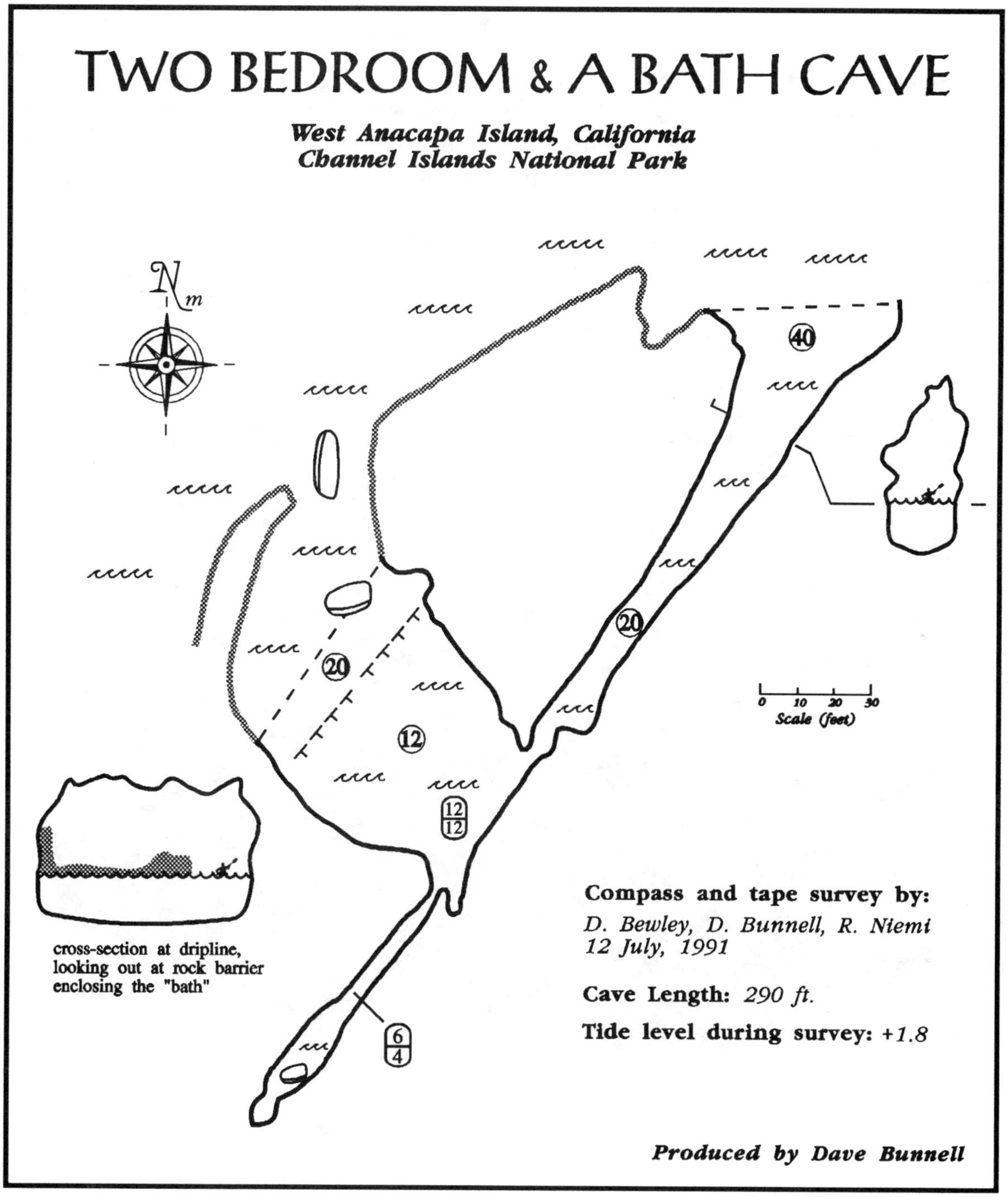
TWO BEDROOM & A BATH CAVE
West Anacapa Island, California
Channel Islands National Park
N
m
40
20
20
12
12
12
6
4
0 10 20 30
Scale (feet)
cross-section at dripline, looking out at rock barrier enclosing the "bath"
Compass and tape survey by:
D. Bewley, D. Bunnell, R. Niemi
12 July, 1991
Cave Length: 290 ft.
Tide level during survey: +1.8
Produced by Dave Bunnell

PURPLE PALACE CAVE - 7

Location: Approx. 3300' from the west tip of West Island

Entrances: 1

Length: 90' (40' for the annex)

Conditions: Mostly dry inside at low tide, so swim to the rocks just in front and explore the cave on foot from there; no lights are needed.

Description: The cave consists of a 35' wide chamber, floored with breakdown and cobbles, with two "arms". The longest arm contains a tidepool and a cobble beach, and is walking-height. The cave was named for the copious red algae on the walls.

Survey crew in Purple Palace Cave

PURPLE PARADISE CAVE

West Anacapa Island, California
Channel Islands National Park

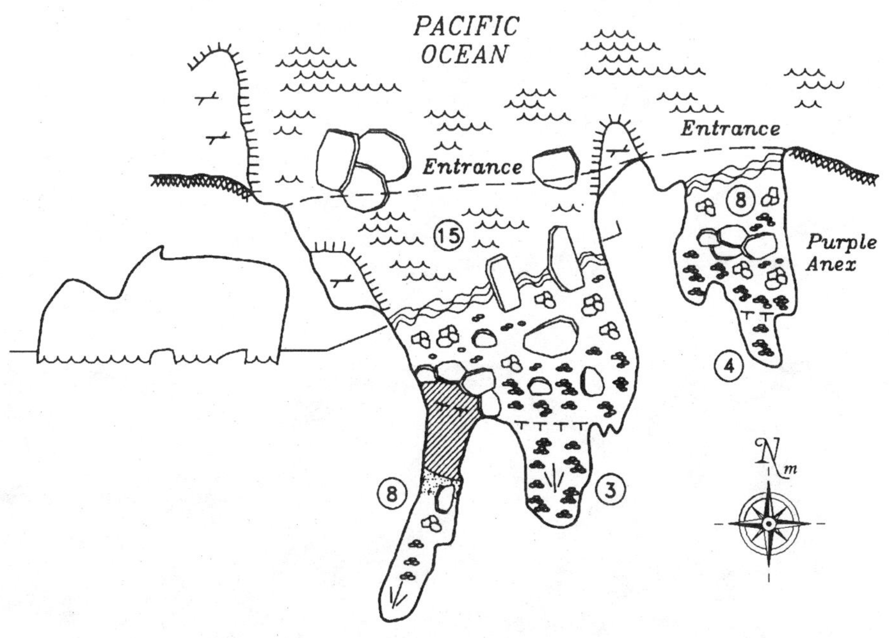

0 10 20 30 40

Scale in Feet

COMPASS and TAPE SURVEY BY:
P. Adams, M. Doe, B. Richards.

DATE: *25 April 1992*
LENGTH: *152 Ft., 46m*
MAP SYMBOLS TO CSCS STANDARDS
AutoCAD Drafting By: *Bob Richards*

TEARDROP CAVE - 8

Location: Approx. 3100' from the west tip of West Island

Entrances: 1

Length: 258'

Conditions: The upper cave is dry and can be explored on foot if one can make a landing on the rocks nearby, problematic unless the seas are calm. There is 12' of exposure as one traverses the fissure to the rear of the cave. The lower level is narrow and tends to amplify swell, and much of it is floored with large urchins. Low tides and calm seas are a necessity to vist this part of the cave. A light is useful for exploring both parts of this cave.

Description: This cave is formed along two distinct levels, like many of the Anacapa caves, with the upper layer in the more easily eroded agglomerate and the lower in the more resistant basalt. The 30' high entrance is shaped like a teardrop. Large breakdown blocks partly obscure the view of the narrow lower passage from the sea. For the first 70' in from the dripline, the cave maintains its teardrop cross-section, with a 12' high fissure in the bottom widening into a broader upper level. After 70' the lower fissure roofs over, producing two distinct passage levels. The lower level continues as a 12' high fissure 3-6' wide and ends in a sloping cobble beach. It is 40' longer than the upper. The upper level can best be reached by climbing up along the cave's right wall. In the floor of the upper level are several skylights which lead down to the lower level. The floor is an irregular basalt. The passage averages 8' high 6' wide and tapers to a narrow end which is heavily encrusted with gypsum.

Kayaking offshore of Teardrop Cave

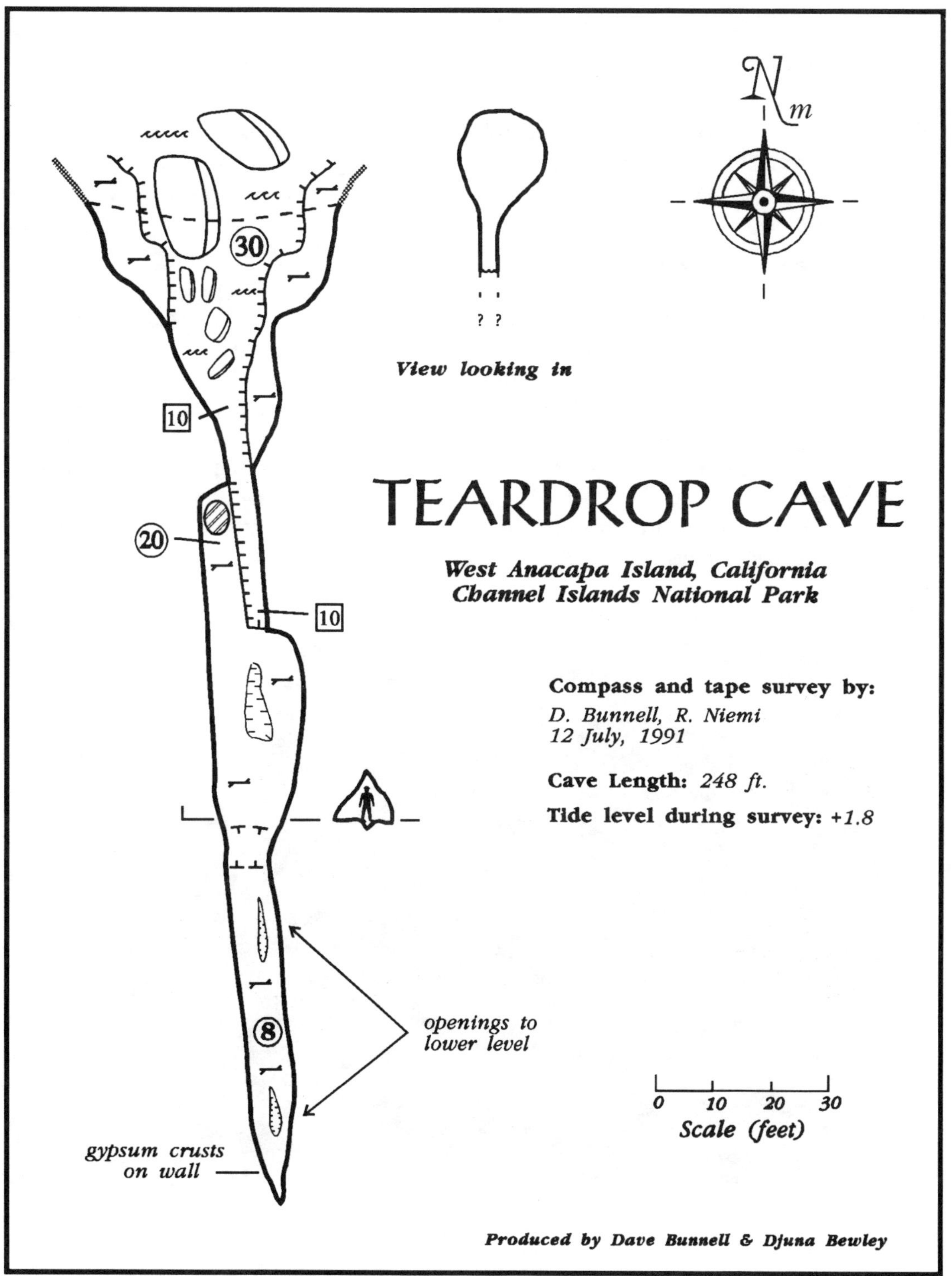

30
10
20
10
8
? ?
View looking in
N
m
TEARDROP CAVE
West Anacapa Island, California
Channel Islands National Park
Compass and tape survey by:
D. Bunnell, R. Niemi
12 July, 1991
Cave Length: 248 ft.
Tide level during survey: +1.8
openings to
lower level
gypsum crusts
on wall
0 10 20 30
Scale (feet)
Produced by Dave Bunnell & Djuna Bewley

THREE FINGERS CAVE - 9

Location: Approx. 5100' from the west tip of West Island.

Entrances: 2

Length: 490'

Conditions: A through-trip between the two entrances may be done in a very short kayak or dinghy if seas are calm. Exploration in the long fingers requires very low tides coupled with very calm seas, and lights. Another finger may extend underwater for an unknown distance.

Description: The eastern entrance is the larger and is 48' wide and 35' high. It leads into a water-filled chamber from which 3 fissures continue. A long pillar separates this chamber from a 15 to 20' high passage extending over 100' to a second entrance. Concerning the three "fingers", the leftmost is 25' long and appears to end; the middle finger is 30' long and appears to continue underwater for an unknown distance. The right-hand "finger" wasn't enterable on our first survey due to the swell. We returned on a calmer day to find it enterable although still intimidating. It continues for over 200' into inky blackness. It varies from 4 to 6' wide and 8 to 15' tall. At the end is a small bedrock-floored chamber about 10' wide, with a small sand beach continuing on the other side. Surprisingly, there are urchins living even here, as the author can attest from the spine that penetrated his wetsuit booties.

Kayaking in Three Fingers Cave

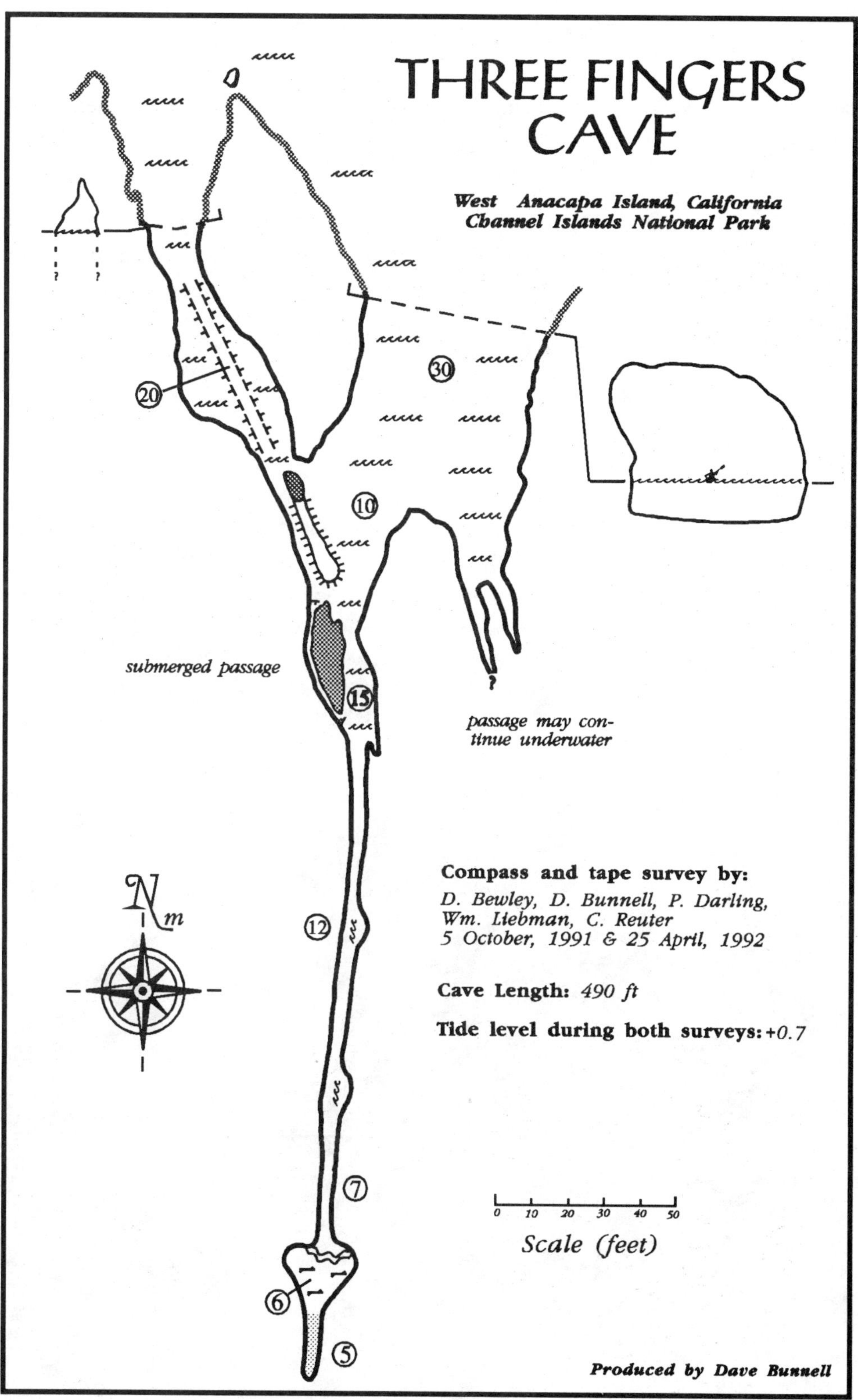
THREE FINGERS CAVE
West Anacapa Island, California
Channel Islands National Park
submerged passage
passage may continue underwater
Compass and tape survey by:
D. Bewley, D. Bunnell, P. Darling,
Wm. Liebman, C. Reuter
5 October, 1991 & 25 April, 1992
Cave Length: 490 ft
Tide level during both surveys: +0.7
0 10 20 30 40 50
Scale (feet)
Produced by Dave Bunnell

IF YOU DARE CAVE - 10

Location: Cave 10 & 11 are approx. 3500' from the west tip of West Island

Entrances: 1

Length: 104'

Conditions: This water-floored cave requires a low tide, calm seas, and a light to explore.

Description: The entrance is 12' high and 20' wide. It leads into a single water-floored passage with some ceilling heights as low as 4', and ends in a solid wall.

TRUTH OR DARE CAVE - 11

Entrances: 1

Length: 170'

Conditions: A water-floored cave which may be explored by kayak or dinghy in higher tides if seas are calm. As the tide lowers, submerged rocks are exposed and the surf breaks on these. No lights are needed. When entering, beware strong currents along the walls.

Description: The cave follows a prominent fault seen along the left wall of the passage. The entrance is some 50' wide and 12' high, and the passage is comfortably wide for the cave's entire length. There are numerous breakdown blocks on the floor of the cave. The cave terminates in a cobble and sand beach. A prominent cobble-floored alcove is seen on the right wall but doesn't seem to lie on a cross-fault.

Truth or Dare (left ent.); If You Dare (right ent)

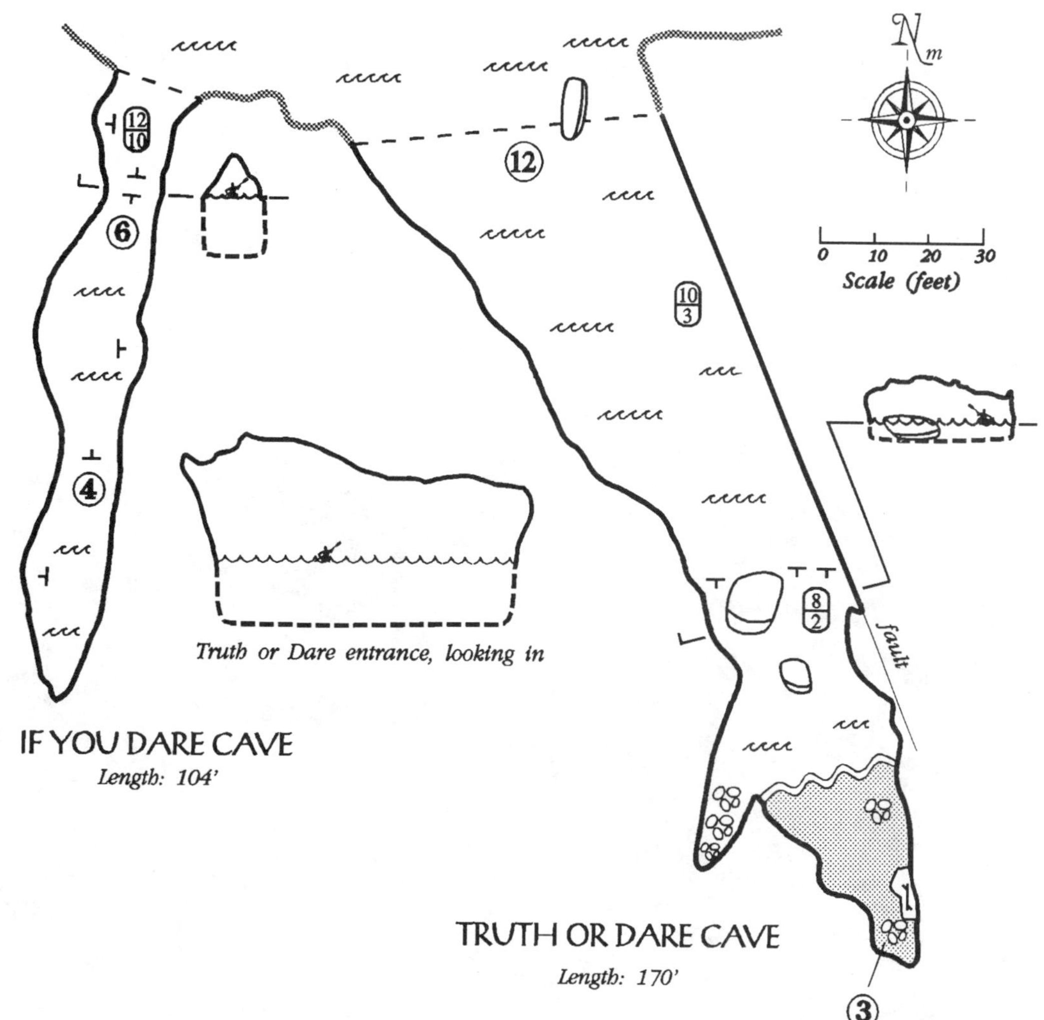

Compass and tape survey by:

D. Bewley, P. Bosted, D. Bunnell, P. Darling, Wm. Liebman
25 April, 1992

Tide level during survey: *+1.1*

Produced by Dave Bunnell

ONE-SHOT CAVE - 12

Location: Approx. 4000' from the west tip of West Island

Entrances: 1

Length: 66'

Conditions: A water-floored cave which may be explored by kayak or dinghy if seas are calm. No lights are needed.

Description: The entrance is 10' high and 40' wide. A large breakdown block is seen just to the left of it. To the right of the entrance are some features which look cave-like but which are very shallow. The cave ends in a very small cobble-covered beach exposed at low tides.

One-Shot Cave on left; unmapped feature on right

ONE-SHOT CAVE

West Anacapa Island, California
Channel Islands National Park

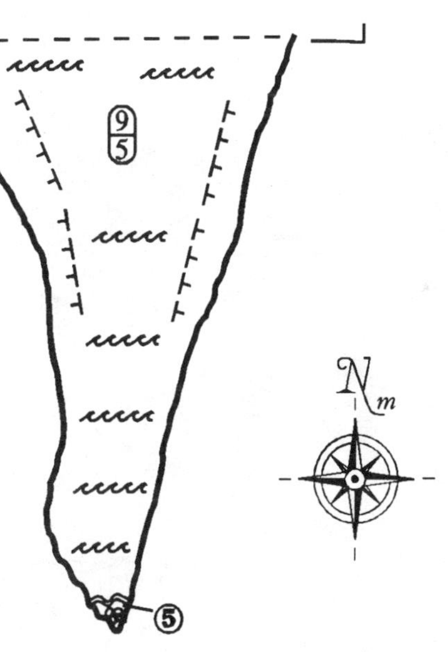

Compass and tape survey by:
D. Bewley, D. Bunnell, C. Reuter
5 October, 1991

Cave Length: *66 ft.*

Tide level during survey: *+2.3*

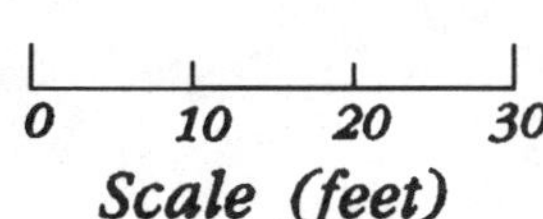

Produced by Dave Bunnell

SLAM-DUNK CAVE - 13

Location: Approx. 4600' from the west tip of West Island

Entrances: 1

Length: 170'

Conditions: This is a cave to visit in only very calm conditions and at a very low tide. Much of it may be submerged at hight tide. Even on a relatively calm day it was very sporting in the back (hence the name). A light is useful as it gets very dark in the back. Not explorable by kayak or dinghy.

Description: At the dripline the cave is some 35' high but necks down after 45' to an arched water-floored passage which continues walking-height for over 100' and ends in a cobble beach.

The less-than-inviting entrance to Slam-Dunk

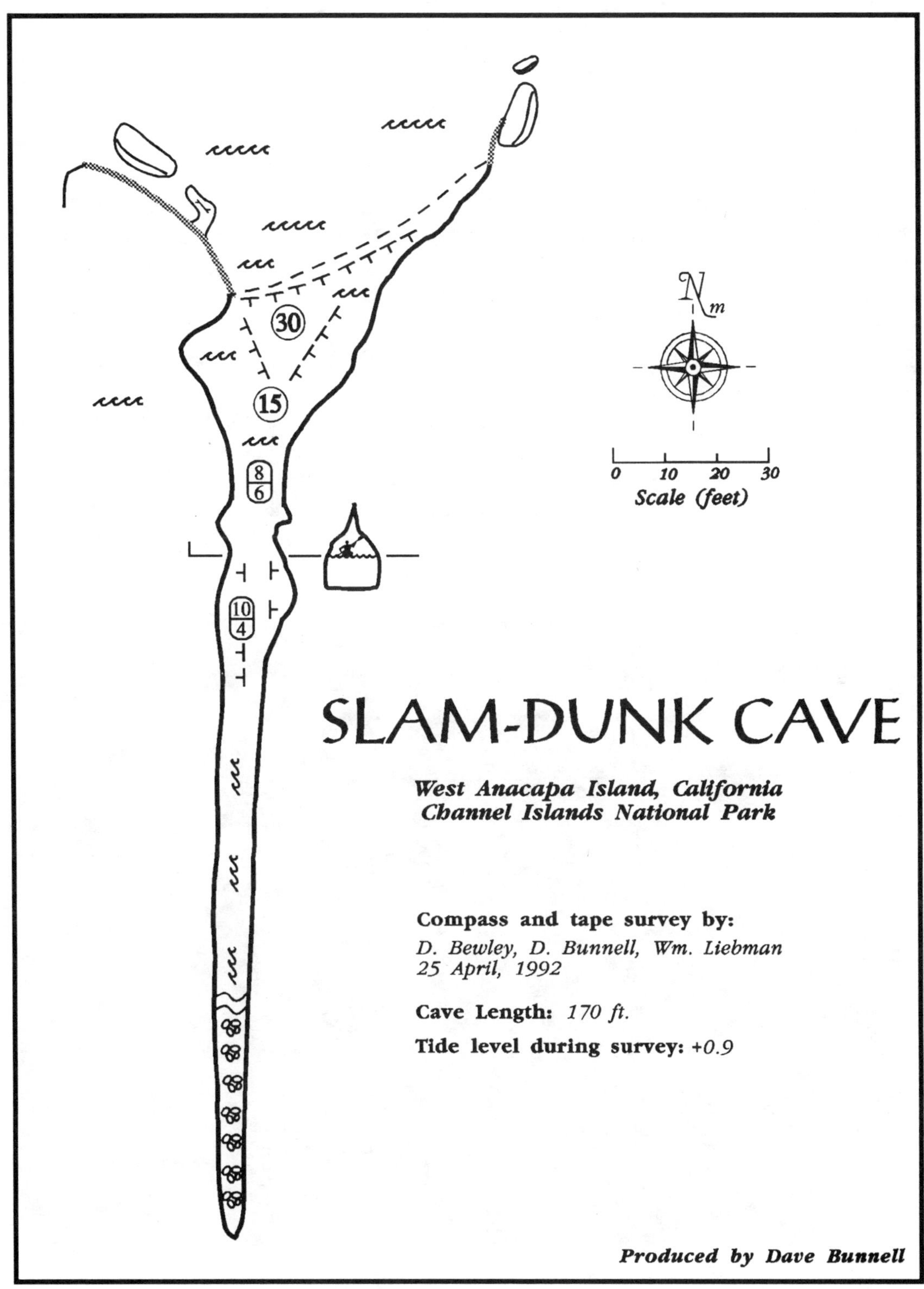
30
15
8
6
10
4
N m
0 10 20 30
Scale (feet)
SLAM-DUNK CAVE
West Anacapa Island, California
Channel Islands National Park
Compass and tape survey by:
D. Bewley, D. Bunnell, Wm. Liebman
25 April, 1992
Cave Length: 170 ft.
Tide level during survey: +0.9
Produced by Dave Bunnell

NESTING CORMORANT CAVE - 14

Location: Approximately 4800' from the west tip of West Island. Marked on the USGS topographic map.

Entrances: 1

Length: 259'

Conditions: Snorkle, swim or kayak in. There is no place to land a kayak but there is room enough to turn around. Avoid entering the cave April though August so as not to disturb the cormorants who nest on ledges above and inside the large entrance. A trip to the rear of the cave should be done only during low tides and calm seas since the passage narrows quite a bit.

Description: The prominent entrance is about 40' high and is 38' wide. The cave follows a fault bearing 175° through large passage which narrows to 3' wide 150' into the cave. Beyond, the passage widens again after it intersects a second fault bearing 210°, along which axis the cave continues another 100' to an end on a sandy beach. Judging by gurgling sounds heard during the survey, the passage may extend out towards the cliff-face along the secondary fault and may contain some airbells separated by submerged passage.

The prominent entrance to Nesting Cormorant Cave

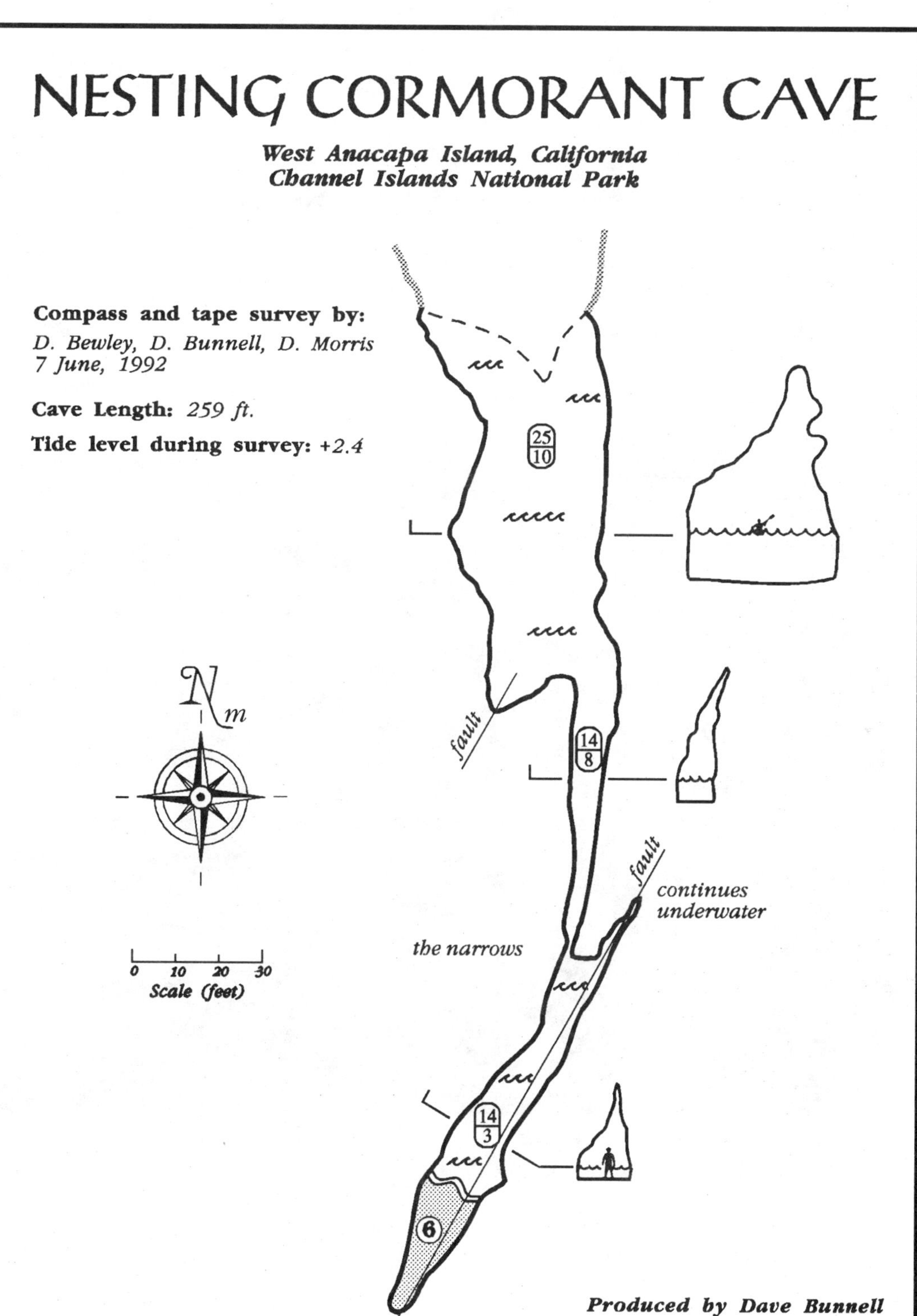
NESTING CORMORANT CAVE
West Anacapa Island, California
Channel Islands National Park
Compass and tape survey by:
D. Bewley, D. Bunnell, D. Morris
7 June, 1992
Cave Length: 259 ft.
Tide level during survey: +2.4
25
10
14
8
14
3
6
fault
fault
continues underwater
the narrows
N
m
0 10 20 30
Scale (feet)
Produced by Dave Bunnell

NO FRILLS CAVE - 15

Location: Approx. 5000' from the west tip of West Island, just west of Sea Lion Cave

Entrances: 1

Length: 41.4'

Conditions: A dry, relict cave. Can be visited at any tide, without a light.

Description: The 20' x 20' entrance is about 10' above sea level. It leads into a small bedrock-floored chamber containing numerous fallen rocks.

Two kayakers paddle past No Frills Cave in a Klepper

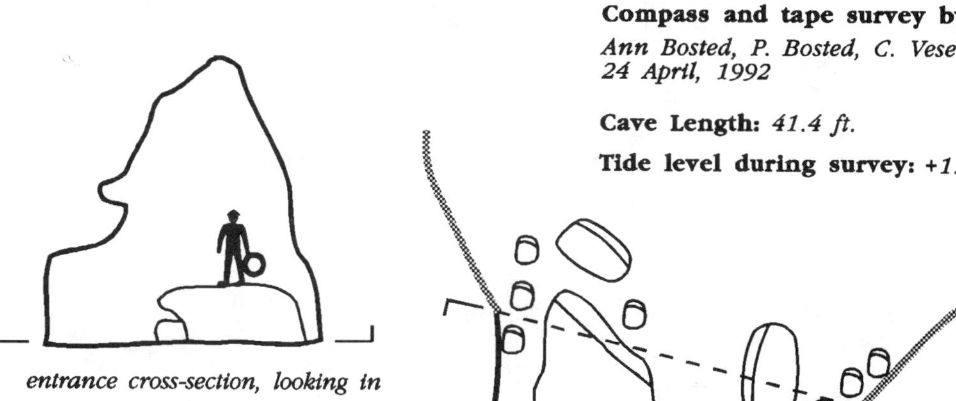
NO FRILLS CAVE
West Anacapa Island, California
Channel Islands National Park
Compass and tape survey by:
Ann Bosted, P. Bosted, C. Vesely
24 April, 1992
Cave Length: 41.4 ft.
Tide level during survey: +1.0
entrance cross-section, looking in
20
15
8
0 10 20 30
Scale (feet)
N
m
Produced by Dave Bunnell

SEA LION CAVE - 16

Location: Approx. 5100' from the west tip of West Island. The two entrances are in a small cove facing NE. Marked on the USGS topographic map.

Entrances: 2

Length: 246' (75 m)

Conditions: The cave is mostly dry at low tide and at times serves as a sea lion habitat, so it is best not to enter this cave when they are present. Submerged rocks in the entrance zone make entry hazardous in heavy surf, but in calm seas a kayak may be landed here. Lights are useful to negotiate the passage connecting the two entrances and to observe the tidepool life. This was the first cave we surveyed on the island.

Description: The cave has two entrances leading into a large cobble-floored chamber which at times houses a large population of California sea lions (perhaps 15-20), and at other times is frequented by harbor seals. Tidepool organisms were limited to the entrance zone and the darker passage connecting the two entrances, where pink-tinted albino anemones were observed. The cave can probably be explored at any tide level provided the surf is low enough to land safely.

Kayaking towards a beach landing in Sea Lion Cave

SEA LION CAVE

West Anacapa Island, California
Channel Islands National Park

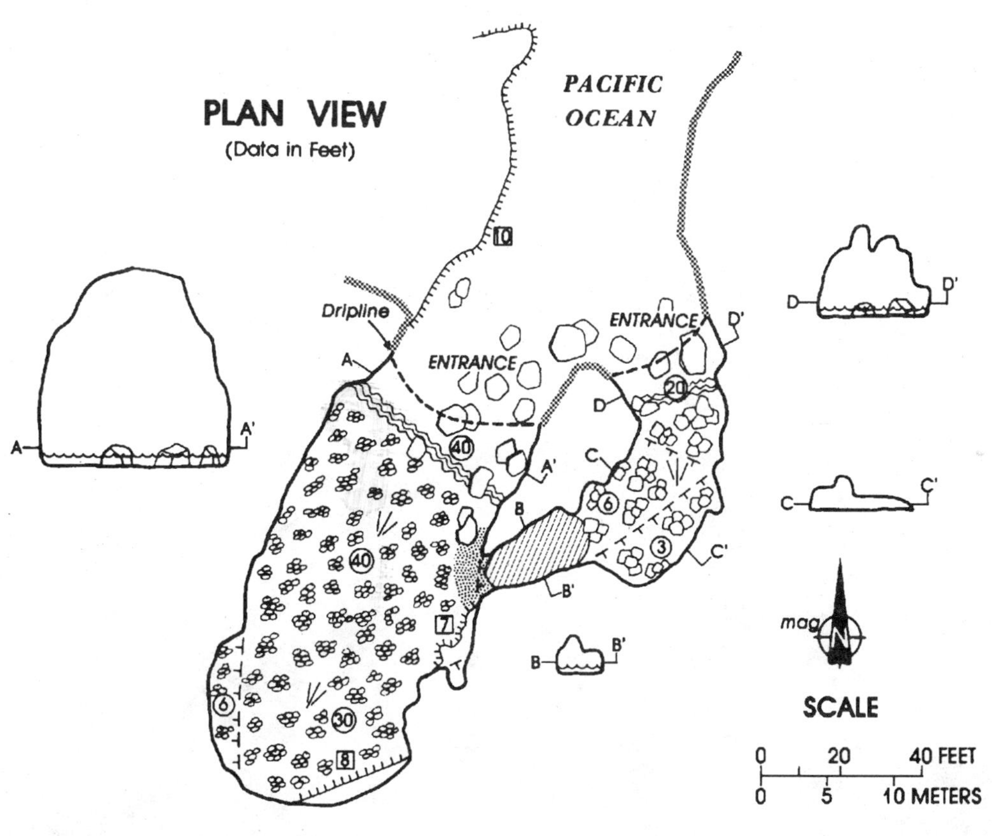

SUUNTO AND TAPE SURVEY:

26 Feb 1984 R. Briesch, D. Bunnell, D. DeLucia, L. DeLucia, B. Richards, C. Vesely.

SURVEYED AT -0.4 Low Tide
TOTAL SURVEYED TRAVERSE: 246 ft, 75 m

CLUB-FOOT CAVE - 17

Location: Approx. 5200' from the west tip of West Island, in the cove just east of Sea Lion Cave

Entrances: 1

Length: 36'

Conditions: Enter during low tide and calm seas. Swim or snorkle in.

Description: A short passage leads into a small, 12' chamber. The cave is entirely underwater. No check was made for submerged passage.

ROUGH GOING CAVE - 18

Entrances: 1

Length: 70'+

Conditions: Enter during low tide and calm seas. Swim or snorkle in. Not explorable by kayak or dinghy.

Description: The entrance is 3' high and 6' wide and leads into a water-floored passage curving to the right. It proved too rough to follow during our survey, and a definitive end wasn't reached.

Club-foot Cave (on right) and Rough Going (on left)

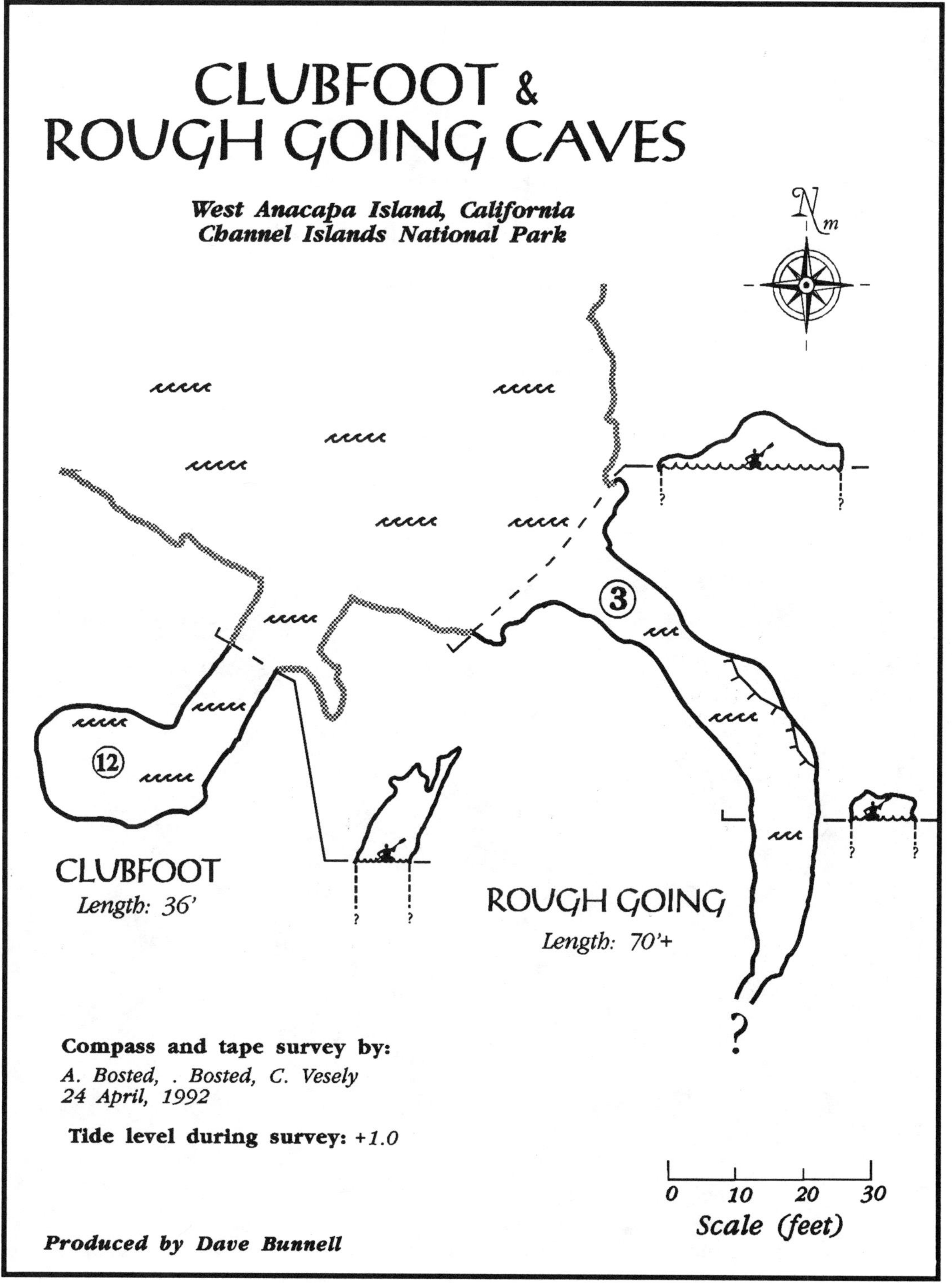
CLUBFOOT &
ROUGH GOING CAVES
West Anacapa Island, California
Channel Islands National Park
N m
3
12
CLUBFOOT
Length: 36'
ROUGH GOING
Length: 70'+
?
Compass and tape survey by:
A. Bosted, . Bosted, C. Vesely
24 April, 1992
Tide level during survey: +1.0
0 10 20 30
Scale (feet)
Produced by Dave Bunnell

GREEN SHELF CAVE - 19

Location: Approx. 5300' from the west tip of West Island. Marked on the USGS topographic map.

Entrances: 1

Length: 158'

Conditions: This cave is best explored by kayak or dinghy. The tide level is relatively unimportant, although a high tide may allow better access to the shelf at the end of the cave. No lights are needed as it is well illuminated through its large entrance.

Description: This is a pleasant little cave floored with very deep water, which could make for an interesting dive. The entrance is 30' wide and 50 to 60' high. The cave maintains its large dimensions until a shelf is reached in the rear. A climb-up to the shelf over red algae gives access to another 40' of cave with a green-algae coated bedrock floor and a couple of tide pools. Ceiling heights drop to 18' and the cave ends in a flat wall (contrasting with the fissures that end most caves).

Kayaker in the rear of Green Shelf Cave

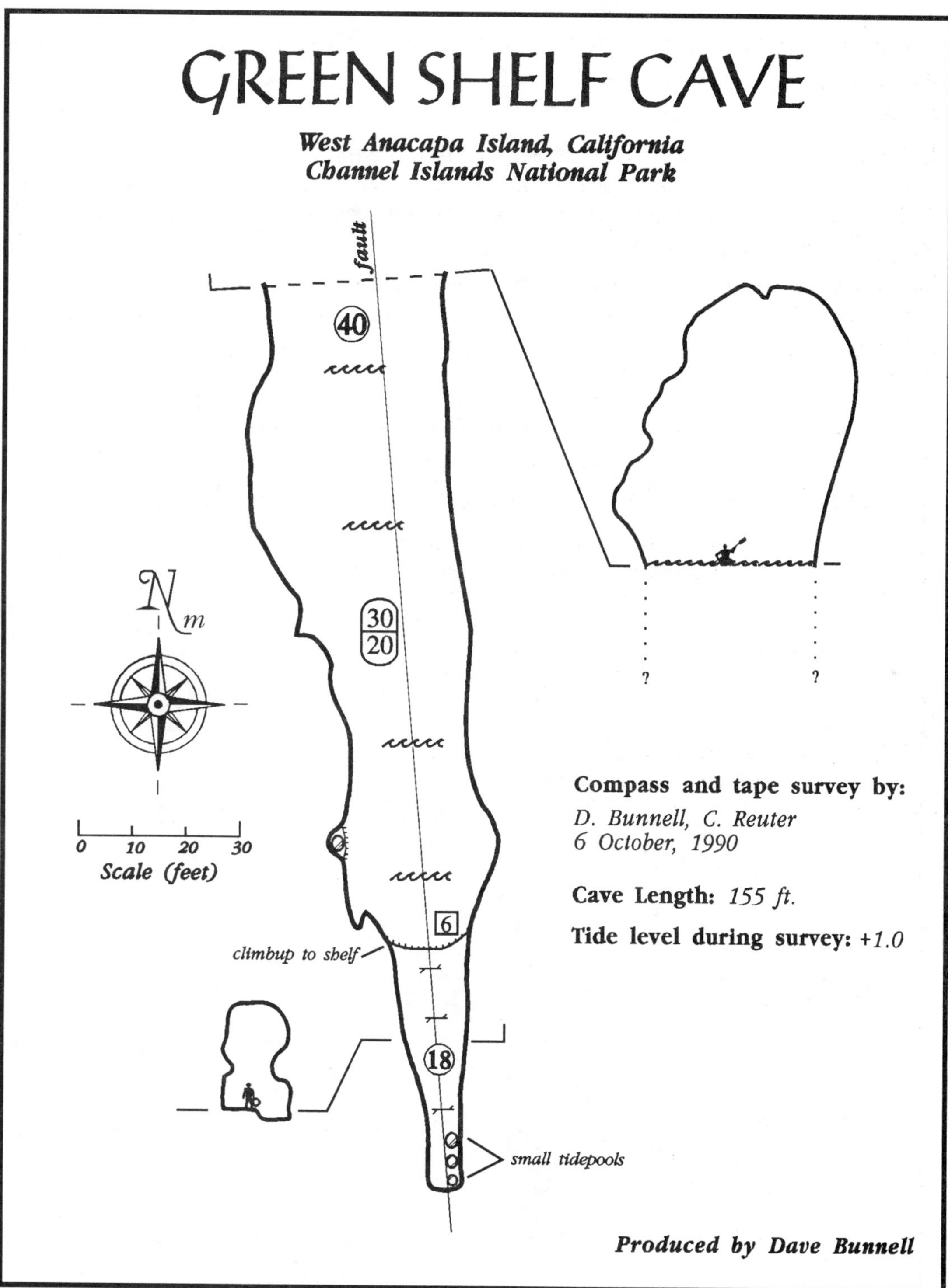
GREEN SHELF CAVE
West Anacapa Island, California
Channel Islands National Park
fault
40
30
20
6
18
N
m
0 10 20 30
Scale (feet)
climbup to shelf
small tidepools
?
?
Compass and tape survey by:
D. Bunnell, C. Reuter
6 October, 1990
Cave Length: 155 ft.
Tide level during survey: +1.0
Produced by Dave Bunnell

SUCKING SLOT CAVE - 20

Location: Approx. 5500' from the west tip of West Island

Entrances: 1

Length: 131'+

Conditions: The portion of the cave shown on the map could likely be visited at any tide level providing seas are calm. A dinghy or kayak could be used to explore it. The "sucking slot" might be enterable on a day with flat seas and negative tides, or possibly with scuba. It could go for some distance...

Description: The cave consists of a single water-floored chamber with a couple of short side passages. The one on the west side is short (23') and ends. The main one has a tricky little climbup to a view into a narrow fissure beyond, which continues an unknown distance.

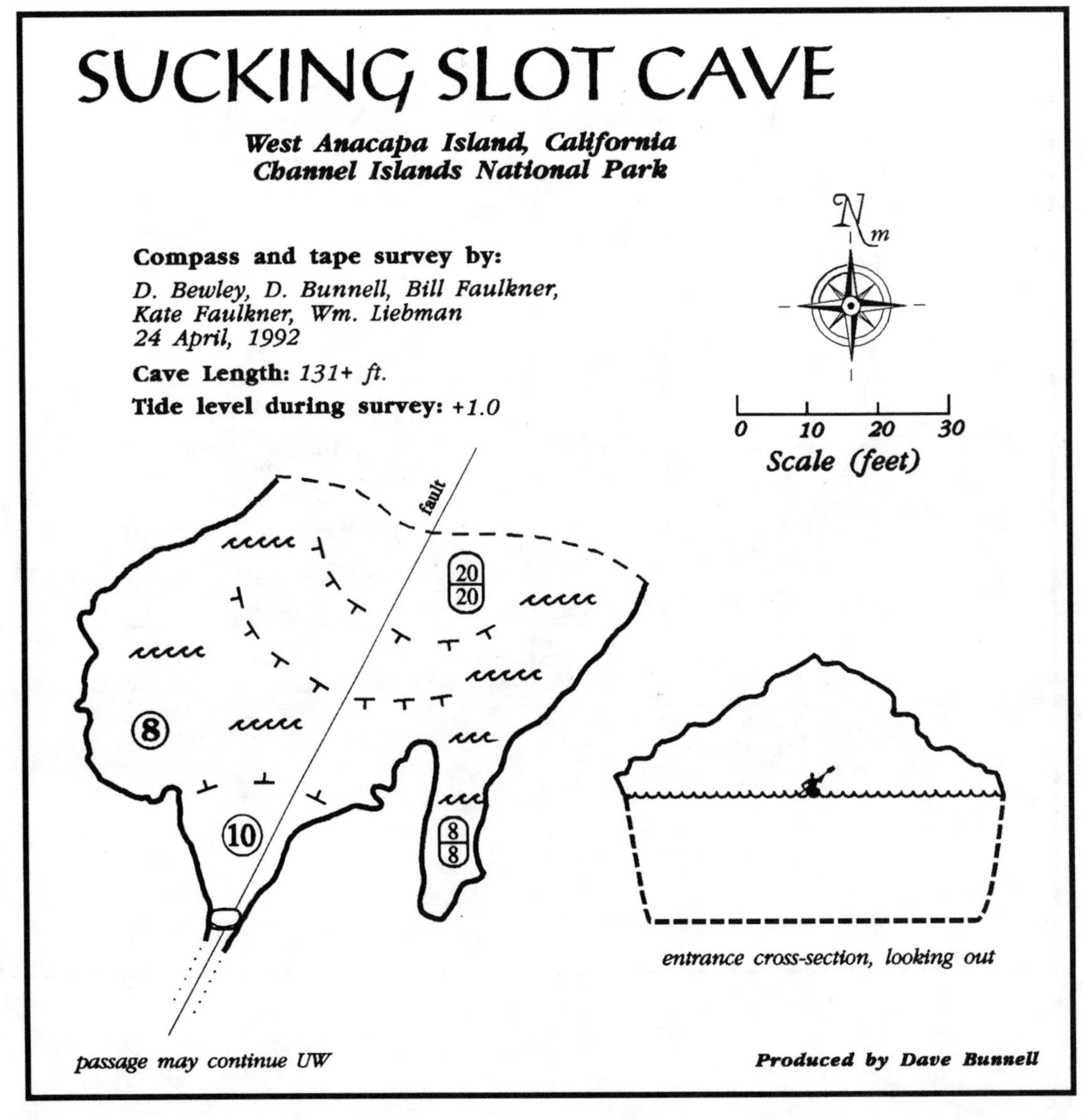

TURBULENCE CAVE - 21

Location: Approx. 5600' from the west tip of West Island

Entrances: 1

Length: 50'

Conditions: Enter during low tide and calm seas. Swim or snorkle in. Not explorable by kayak or dinghy. It is difficult to stay in any one spot in here, with the swell producing currents that wrap around the pillar inside.

Description: This cave consists of a small water-floored chamber with a pillar dividing it. There don't appear to be any underwater passages extending from the room.

Sea caver heads into Turbulence Cave, now calm

TRANQUILITY CAVE - 22

Location: Approx. 5600' from the west tip of West Island

Entrances: 1

Length: 114'+

Conditions: Most of the cave is high and dry, so one can explore it on foot at any tide level.

Description: The cave is formed along two parallel faults trending due south. The cave is floored with bedrock and ceiling heights vary from 6 to 12'. A submerged passage is visible along the left-hand fault, but it was not explored.

Survey crew taking tape reading in Tranquility Cave

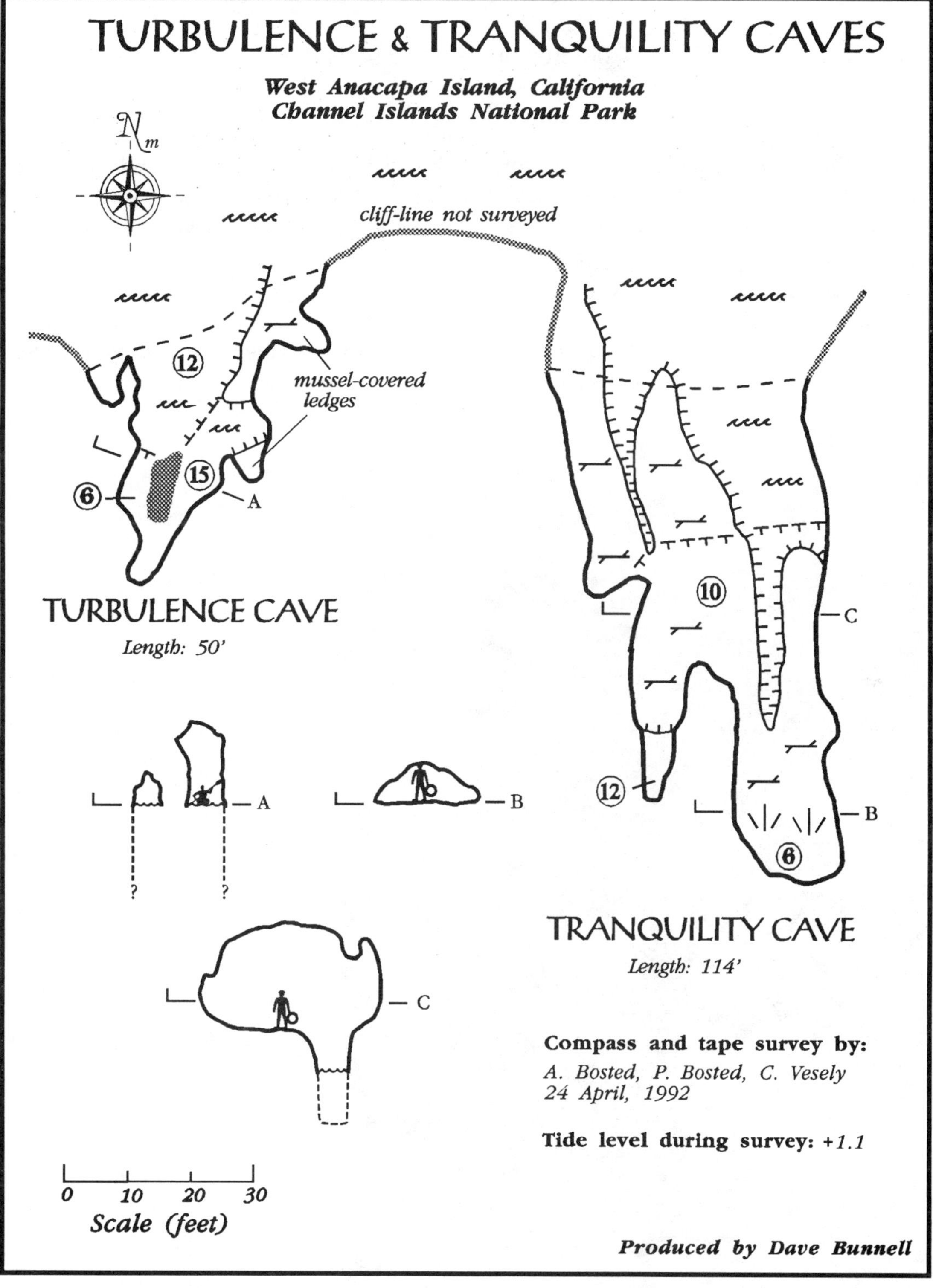
TURBULENCE & TRANQUILITY CAVES
West Anacapa Island, California
Channel Islands National Park
cliff-line not surveyed
mussel-covered ledges
TURBULENCE CAVE
Length: 50'
TRANQUILITY CAVE
Length: 114'
Compass and tape survey by:
A. Bosted, P. Bosted, C. Vesely
24 April, 1992
Tide level during survey: +1.1
0 10 20 30
Scale (feet)
Produced by Dave Bunnell

WOODEN LETTUCE CAVE - 23

Location: Approximately 5700' from the west tip of West Island

Entrances: 1

Length: 63'

Conditions: Swim or snorkle in. A short dinghy might enter and turn around. The cave seems to continue underwater for an unknown distance. Best avoided April though August to prevent disturbing the cormorants who nest on ledges above and inside the entrance, if they are present.

Description: The entrance is about 20' high and 10' wide. The passage widens to about 15' wide inside. At the rear the cave necks down to a narrow fissure, through which one can view some 20' or so of airspace above the water. Below water, the passage is wider and continues an unknown distance.

The entrance to Wooden Lettuce

WOODEN LETTUCE CAVE

West Anacapa Island, California
Channel Islands National Park

Compass and tape survey by:
D. Morris (solo survey)
24 April, 1992

Cave Length: *63+ ft.*

Tide level during survey: *+2.0*

N m

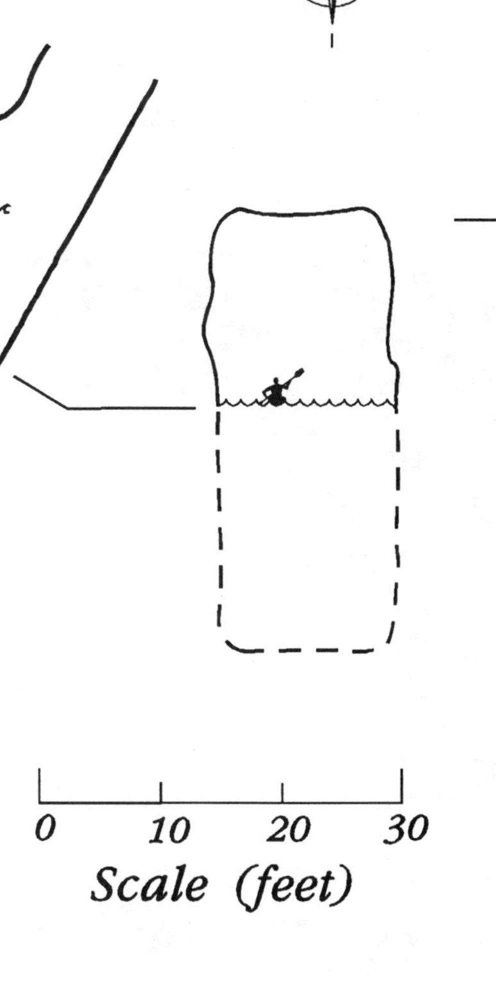

15

passage continues, submerged

Produced by Dave Bunnell

ISLAND VIEW CAVE -24

Location: In the west end of a small cove, 4800' west of Frenchy's Cove. Marked on the USGS topographic map. The cave is in the pelican closure area and may currently be visited only from November 1-Dec. 31 (but check current regulations).

Entrances: 1

Length: 272'

Conditions: This cave could be probably explored at any tide level, although at lower tide a dinghy could be landed in the rear. There are numerous submerged rocks, and the clearest landing would probably be on the far right. Lights are needed in the rear portions.

Description: The entrance is some 40' high and 20' wide but quickly lowers to 15' high as it leads into a spacious tunnel which curves to the left and widens into a larger room. There is a long cobble beach in the rear with three separate "fingers" and a large pillar. Ceiling heights are eight to 20 feet throughout the cave. There is some minor dripstone and some gypsum deposits. The cave is named for the view of Middle and East Anacapa afforded when looking out the east-facing entrance. This is likely the cave visited by Yates (1891), who termed it "Dark Cave".

Looking out from Island View Cave. East Anacapa in distance.

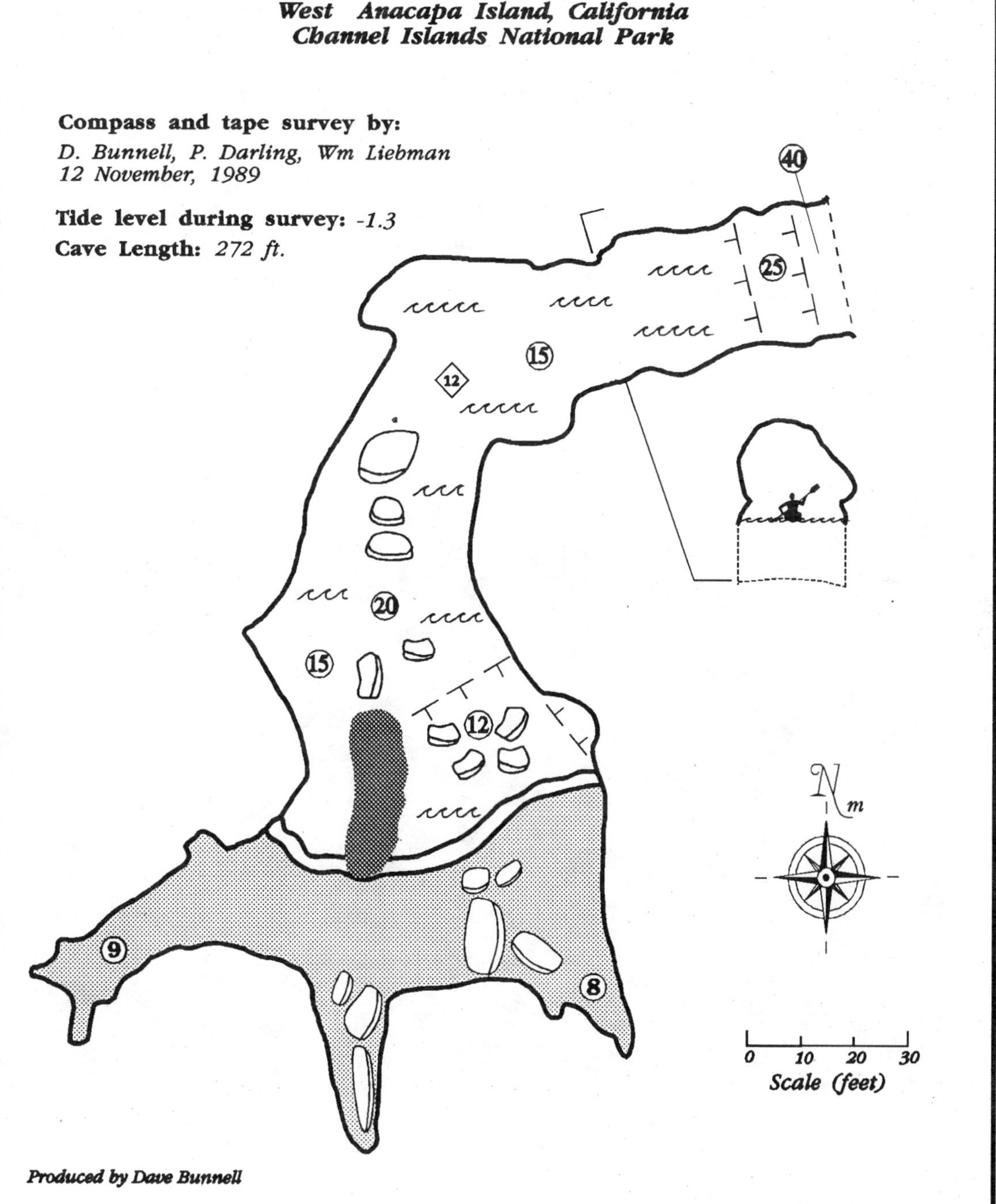
ISLAND VIEW CAVE
West Anacapa Island, California
Channel Islands National Park
Compass and tape survey by:
D. Bunnell, P. Darling, Wm Liebman
12 November, 1989
Tide level during survey: -1.3
Cave Length: 272 ft.
40
25
15
12
20
15
12
9
8
m
0
10
20
30
Scale (feet)
Produced by Dave Bunnell

CONFUSION CAVE - 25

Location: In the middle of a small cove, 4800' west of Frenchy's Cove. The cave is in the pelican closure area and may currently be visited only from November 1-Dec. 31 (but check current regulations).

Entrances: 1

Length: 242'

Conditions: The outer portions of the cave may be explored by dinghy. The back portions require low tides and calm seas to explore, by swimming and wading.

Description: This cave has an impressive arched entrance, some 60' wide and 70' high. Beyond the large entrance chamber, the cave splits into two fingers which are joined at the rear by a low passage at low tide. The lefthand finger is divided by ceiling pendants into two passages, the leftmost of which extends to a small sand beach. The righthand finger extends into an alcove which takes the larger force of the west swell.

A view into Confusion cave during calm sea conditions

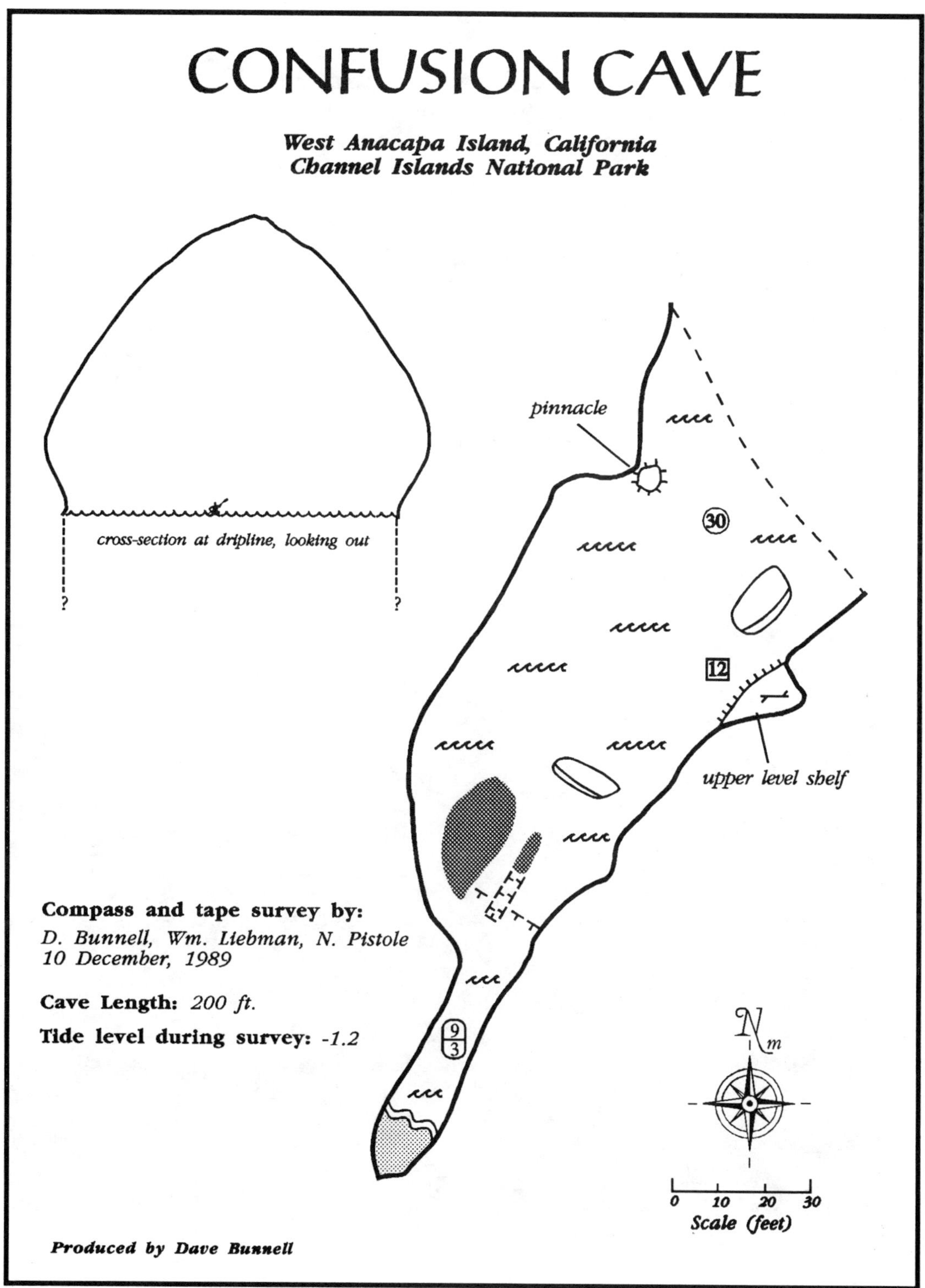
CONFUSION CAVE
West Anacapa Island, California
Channel Islands National Park
cross-section at dripline, looking out
?
?
pinnacle
30
12
upper level shelf
9
3
Compass and tape survey by:
D. Bunnell, Wm. Liebman, N. Pistole
10 December, 1989
Cave Length: 200 ft.
Tide level during survey: -1.2
N m
0 10 20 30
Scale (feet)
Produced by Dave Bunnell

PINNACLE CAVE - 26

Location: In the east end of a small cove, 4800' west of Frenchy's Cove. Marked on the USGS topographic map. The cave is in the pelican closure area and may currently be visited only from November 1-Dec. 31 (but check current regulations).

Entrances: 1

Length: 164'

Conditions: The cave itself is high and dry. There is no real good place to land a dinghy unless the seas are calm enough to pull it up on the rocks below the entrance.

Description: The cave is named for a large pinnacle that juts up from the large pile of fallen rocks at the cave's entrance. The latter is some 50' wide and perhaps 60' high. Some 30' higher than the roof of this entrance is a relict sea cave which is unreachable from below. The main cave is a single passage floored with small broken rock. There was little of note in the cave other than some bird bones in the rear. Erosion of a passage at current sea level in Pinnacle Cave has been prevented by the collapse at the entrance.

Looking in to Pinnacle Cave (center, behind large block); the pinnacle is on the left

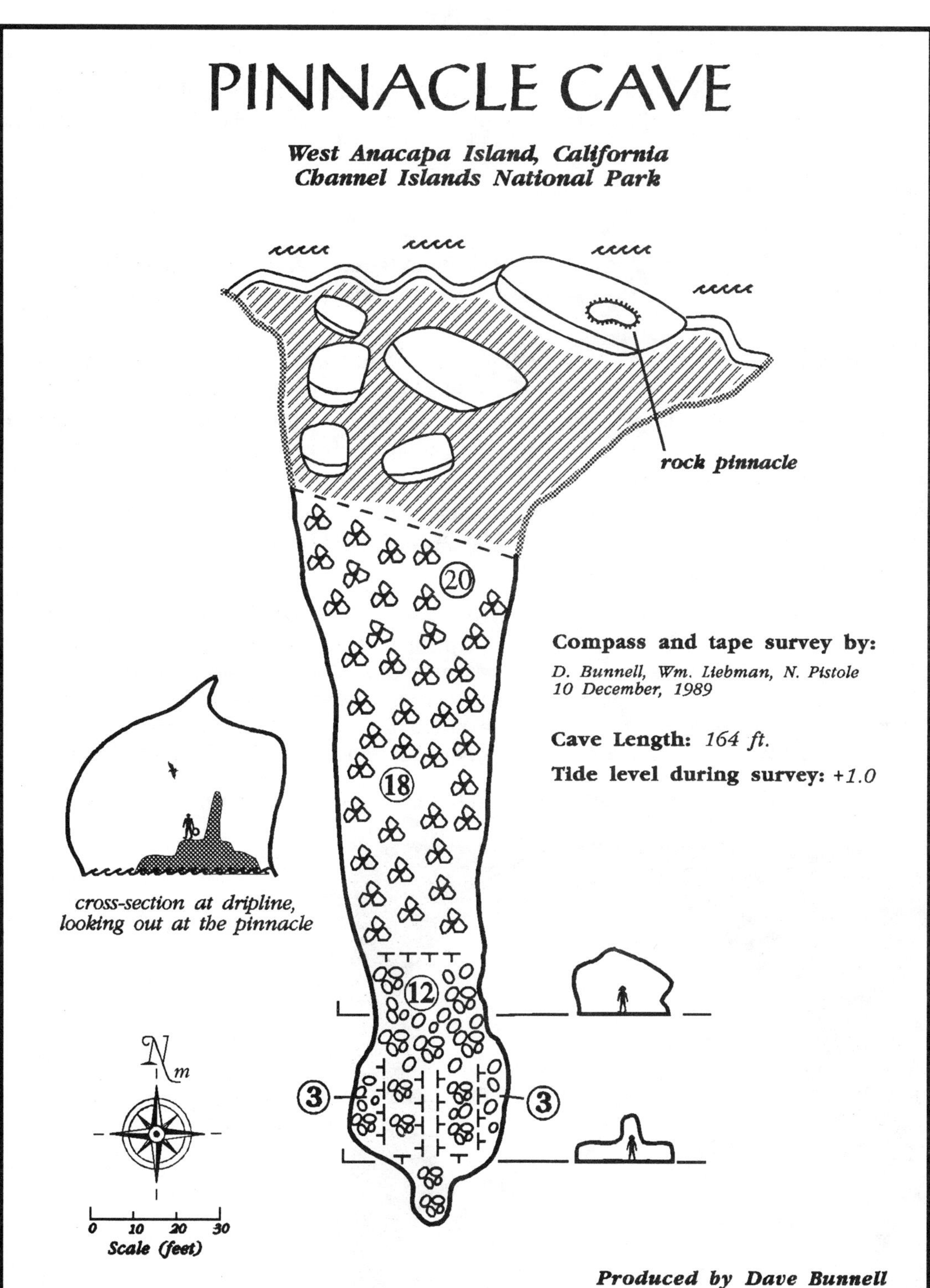
PINNACLE CAVE
West Anacapa Island, California
Channel Islands National Park
rock pinnacle
20
18
12
3
3
Compass and tape survey by:
D. Bunnell, Wm. Liebman, N. Pistole
10 December, 1989
Cave Length: 164 ft.
Tide level during survey: +1.0
cross-section at dripline, looking out at the pinnacle
N m
0 10 20 30
Scale (feet)
Produced by Dave Bunnell

MOSS CAVE - 27

Location: About 300' west of Frenchy's Cave.The cave is in the pelican closure area and may currently be visited only from November 1-Dec. 31 (but check current regulations).

Entrances: 1

Length: 100'

Conditions: The cave itself is completely dry but requires either a swim to shore or sea conditions calm enough to permit a kayak/dinghy landing amongst the rocks in front of the cave.

Description: The cave has a 50' high and 35' wide entrance affording a nice view of Middle and East Anacapa. Large breakdown blocks dominate the area below the dripline, past which the floor is largely bare bedrock for 30'. On the left wall in this area a fresh-water seep emerges from a fault and fills a small pool on the floor. The wall is covered with a carpet of moss fed by this water. Beyond this point the ceiling lowers gradually and the floor is covered with angular broken rocks. This cave is likely one of those described by Yates in his Jan. 1890 article in *The American Geologist*, where he called it Freshwater or Indian cave. He states:

> *At the mouth of this cave is a spring of good water seeping from the rocks into basin-shaped cavities which are evidently artificial. One of these fills up at the rate of 70 gallons every 24 hours. Among the refuse matter deposited in this cave by the Indians we found but little except some fragments of ropes made of sea-grass. Some of these ropes were braided with three strands, the others twisted like ordinary rope used at the present day. We also found bones of a variety of animals which had been used as food.*

There is no evidence of Indian use visible in the cave today, however.

Looking out of Moss Cave towards East and Middle Islands

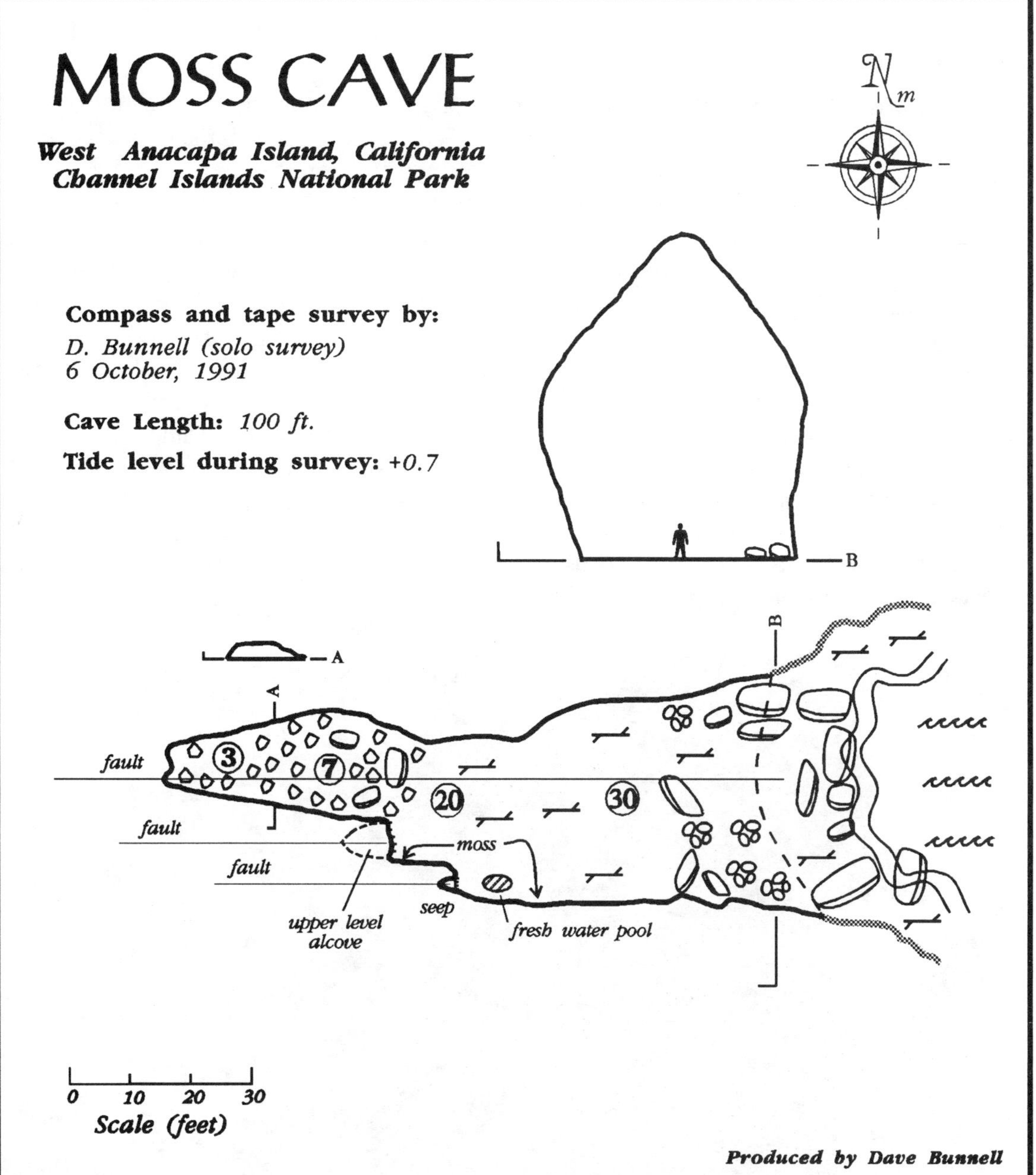
MOSS CAVE
West Anacapa Island, California
Channel Islands National Park
Compass and tape survey by:
D. Bunnell (solo survey)
6 October, 1991
Cave Length: 100 ft.
Tide level during survey: +0.7
A
B
fault
fault
fault
3
7
20
30
moss
seep
upper level
alcove
fresh water pool
0 10 20 30
Scale (feet)
Produced by Dave Bunnell

FRENCHY'S CAVE - 28

Location: About 2000' west of Frenchy's Cove. Marked on the USGS topographic map. The cave is in the pelican closure area and may currently be visited only from November 1-Dec. 31 (but check current regulations).

Entrances: 1

Length: 421'

Conditions: This cave can be explored by dinghy or kayak at any but a very low tide level, when numerous submerged rocks are exposed. The cave is generally well sheltered from swell and in relatively calm conditions a landing may be made inside on either of the two ends of the room. No lights are needed to navigate within the cave.

Description: This cave is also known as Yates Cave, having been first described by geologist Lorenzo Yates in 1890 (see page 7). Later it was used by island resident Frenchy LaDreau for storing bootleg liquor during the Prohibition era. It is one of the most spacious and interesting of the caves on the island. The cave is "T"-shaped, with a short passage following a fault intersecting an impressive chamber some 200'long, 100' wide, and 50-60' high. At each end of the chamber is a cobble beach. Along the entire back wall of the chamber is an unusual sand slope which extends steeply to the ceiling. This slope is well above normal tidal range and its origin is unknown. A fissure passage following the trend of the entrance-passage fault continues into the bedrock underlying the sand slope. Gypsum crusts line the walls at the edges of the chamber, some colored by green algal growths. Seals don't seem to use this cave, but there is abundant tidepool life including the ever-present purple urchins.

Paddling in to Frenchy's Cave

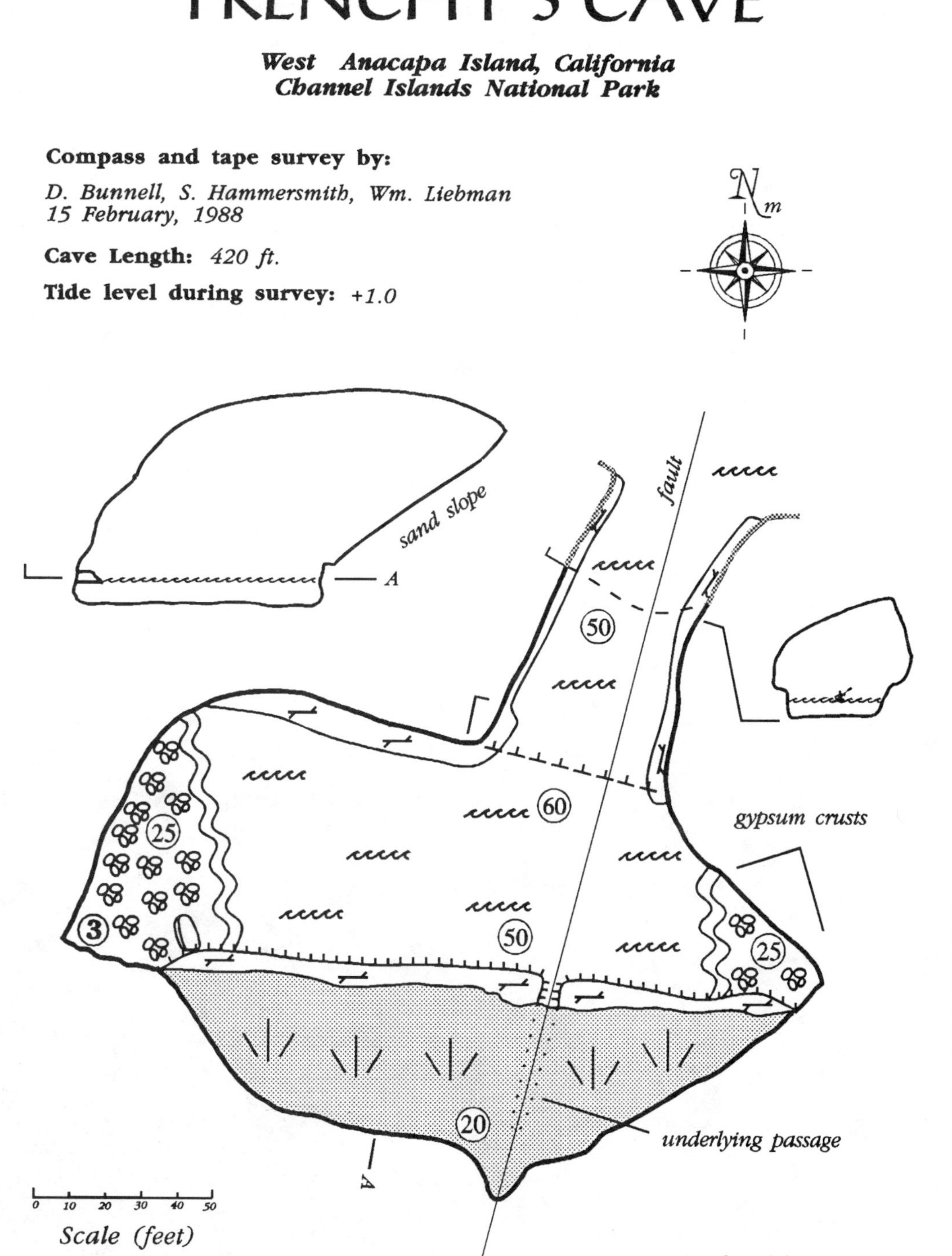
FRENCHY'S CAVE
West Anacapa Island, California
Channel Islands National Park
Compass and tape survey by:
D. Bunnell, S. Hammersmith, Wm. Liebman
15 February, 1988
Cave Length: 420 ft.
Tide level during survey: +1.0
N m
sand slope
A
fault
50
60
25
3
50
25
20
gypsum crusts
underlying passage
A
0 10 20 30 40 50
Scale (feet)
Produced by Dave Bunnell

FRENCHY'S COVE CAVE - 29

Location: Just east of the surge channel dividing West Island and the large stack between West and Middle islands.

Entrances: 1

Length: 62'

Conditions: Empties out at low tide, leaving a sandy beach. No light is necessary.

Description: This is a single passage cave with an entrance 6' high and 4' wide. Ceiling heights inside are generally walking or stooping height. The floor is all sand. A large piece of rusting metal wreckage of undetermined age was seen, partially buried in the sand.

FRENCHY'S TWIN ARCH - 30

Location: Fronts the small cove and beach on the north shore of the large sea stack separating Middle and West Islands.

Entrances: 3

Length: 77'

Conditions: One can traverse these twin tunnels at low tide if the seas are calm. No lights are needed.

Description: This cave consists of a pair of walking-height tunnels, one averaging about 10' wide and 15-20' high and the other much narrower, perhaps 3-4' wide throughout.

Frenchy's Cove Cave on right; Twin Arch on left

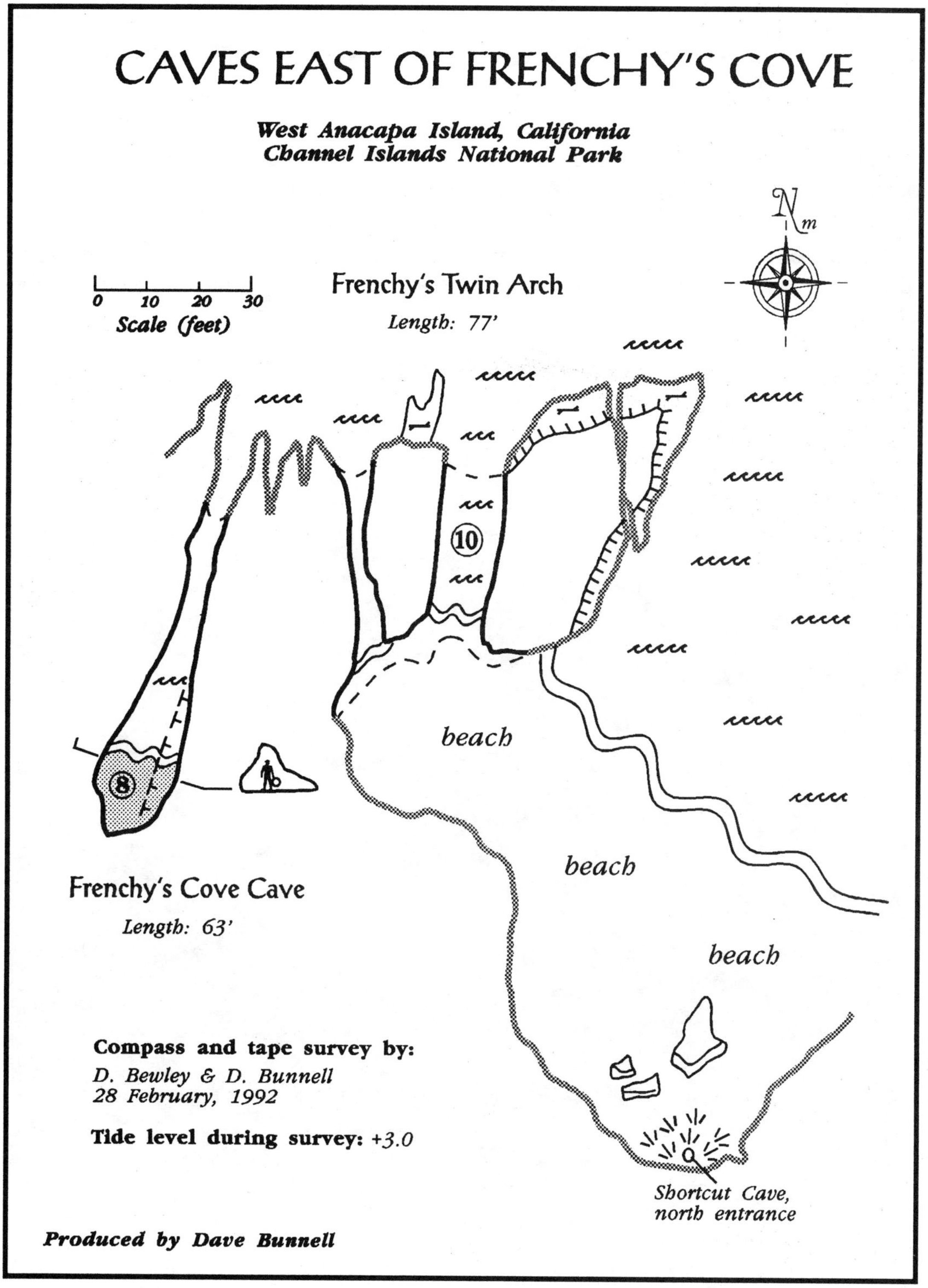

CAVES EAST OF FRENCHY'S COVE

West Anacapa Island, California
Channel Islands National Park

Nm

0 10 20 30
Scale (feet)

Frenchy's Twin Arch

Length: 77'

Frenchy's Cove Cave

Length: 63'

beach

beach

beach

Shortcut Cave, north entrance

Compass and tape survey by:
D. Bewley & D. Bunnell
28 February, 1992

Tide level during survey: *+3.0*

Produced by Dave Bunnell

LEAPYEAR CAVE - 31

Location: on the north shore of the large sea stack separating Middle and West Islands.

Entrances: 1

Length: 85+' (lower level not surveyed)

Conditions: The lower level would require a low tide & calm seas to explore. The upper level could be visited at any tide level or sea condition provided one can safely land on the rocks outside. No lights are needed.

Description: The cave has an impressive entrance, some 25' high and 50' wide at the dripline. The outer portion of the cave is floored with bedrock, while the inner portions are covered by a steep sand slope, with a 35-40° angle of repose. Numerous bird feathers cover the slope.

Leapyear Cave, with dry upper level and wet lower level

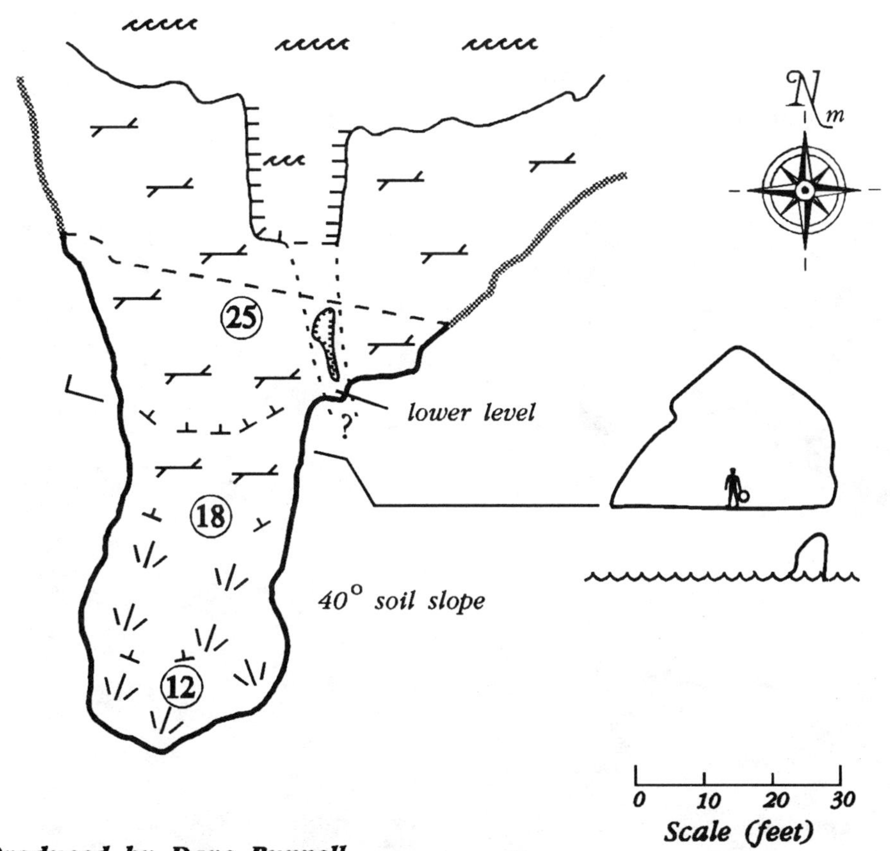
LEAPYEAR CAVE
West Anacapa Island, California
Channel Islands National Park
Compass and tape survey by:
D. Bewley & D. Bunnell
29 February, 1992
Cave Length: 85+ ft.
Tide level during survey: +3.0
N m
25
18
12
?
lower level
40° soil slope
0 10 20 30
Scale (feet)
Produced by Dave Bunnell

LONG BEACH CAVE - 32

Location: On the south shore of the larger of the two islets between West and Middle islands. The second, smaller entrance isn't visible from a seaward approach.

Length: 372'

Entrances: 2

Conditions: The cave is best visited at low tide when seas are calm, although the ceilings are high enough to permit higher tide visits as well. The cave requires swimming, wading, and walking to fully explore but is suitable for visitation by dinghy or kayak, conditions permitting. A light is useful for exploring the side passage and the rear of the main passage.

Description: The cave is formed along two axes. The main entrance is 15' wide and 18' high and leads into a main passage which extends over 200'. At the extreme low tide during our survey, over half this passage was floored with fine sand. The passage width varies from 20 to 35' and ceiling heights average 10'. The walls in this passage are smoothly sculpted. Some 25' in from the main entrance the second axis is intersected. To the right, the passage extends 50' to an entrance 8' wide and 12' high in a surge channel which parallels the cliffs. To the left, this passage continues over 100' into darkness, ending in a sloping cobble beach.

Main entrance of Long Beach Cave, at low tide

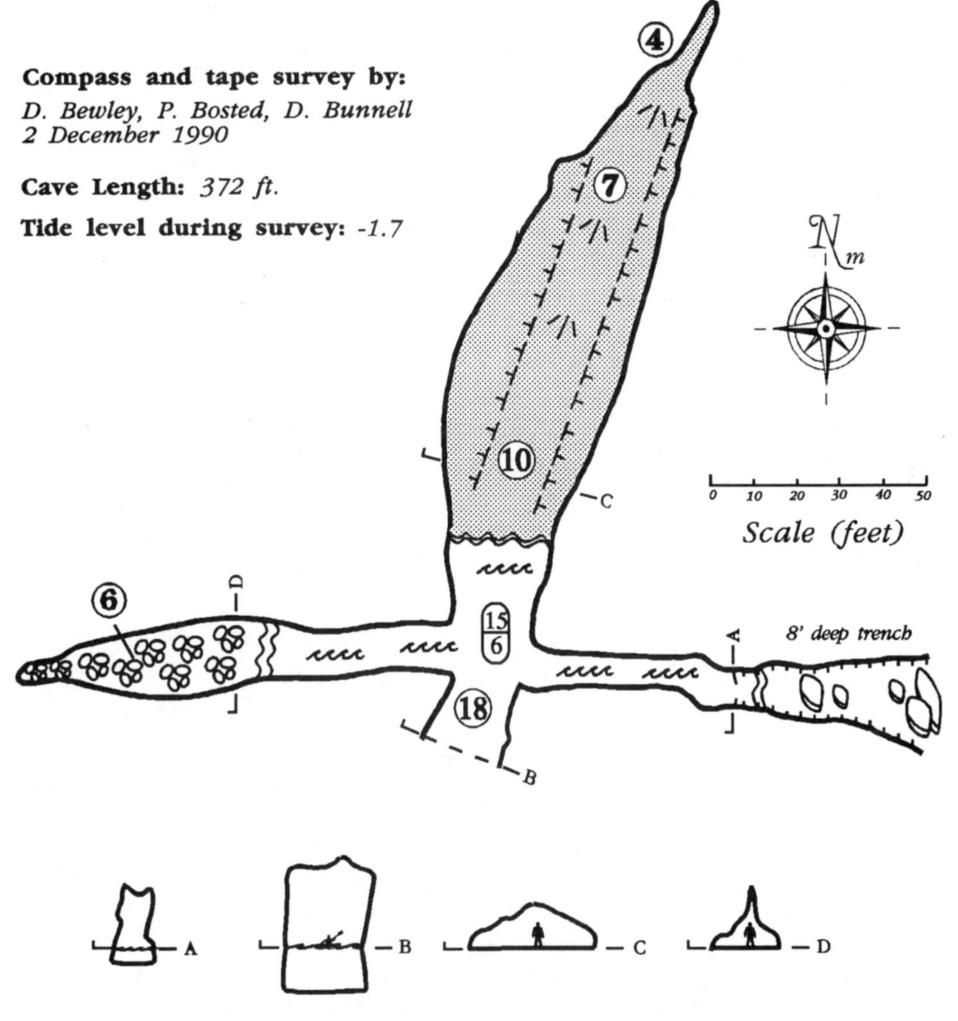
LONG BEACH CAVE
West Anacapa Island, California
Channel Islands National Park
Compass and tape survey by:
D. Bewley, P. Bosted, D. Bunnell
2 December 1990
Cave Length: 372 ft.
Tide level during survey: -1.7
4
7
10
15
6
18
6
8' deep trench
A
B
C
D
N
m
0 10 20 30 40 50
Scale (feet)
Produced by Dave Bunnell

SHORTCUT CAVE - 33

Location: The northern (dry) entrance is on the small beach on the north side of the large stack separating East and Middle islands, and isn't visible until one is right on top of it. There are two larger entrances on the south shore of the stack.

Entrances: 3

Length: 164'

Conditions: The northern entrance is high and dry; the southern entrance is open at low tide but requires a swim to reach, except at a very low tide when access is possible from Frenchy's Cove. No lights are necessary to do the through-trip but crawling is required.

Description: This unusual cave cuts completely through this small islet, here only 82' wide. Approaching from the south, the left-hand entrance leads into a chamber 50' long, 25' wide and up to 6' high with a sloping cobble floor. A 25' long passage on the right wall connects through (at low tide) to another smaller entrance and a dry tunnel to the left. This tunnel is some 5-6' tall and carries a tell-tale breeze from the northern entrance, which is not visible until one turns a corner midway through the passage. The passage then slopes up and emerges via a crawlway over finely broken rock.

SHORTY CAVE - 34

Location: On the south shore, just east of the surge channel dividing West Island and the large stack between East and Middle islands.

Entrances: 1

Length: 43'

Conditions: Visit at low tide only, when it empties out, and access by foot is possible from Frenchy's Cove.

Description: The cave is a small chamber floored with unusually large cobbles. Most of the chamber is lower than walking-height.

Shorty Cave (left); Shortcut Cave, south entrance (right)

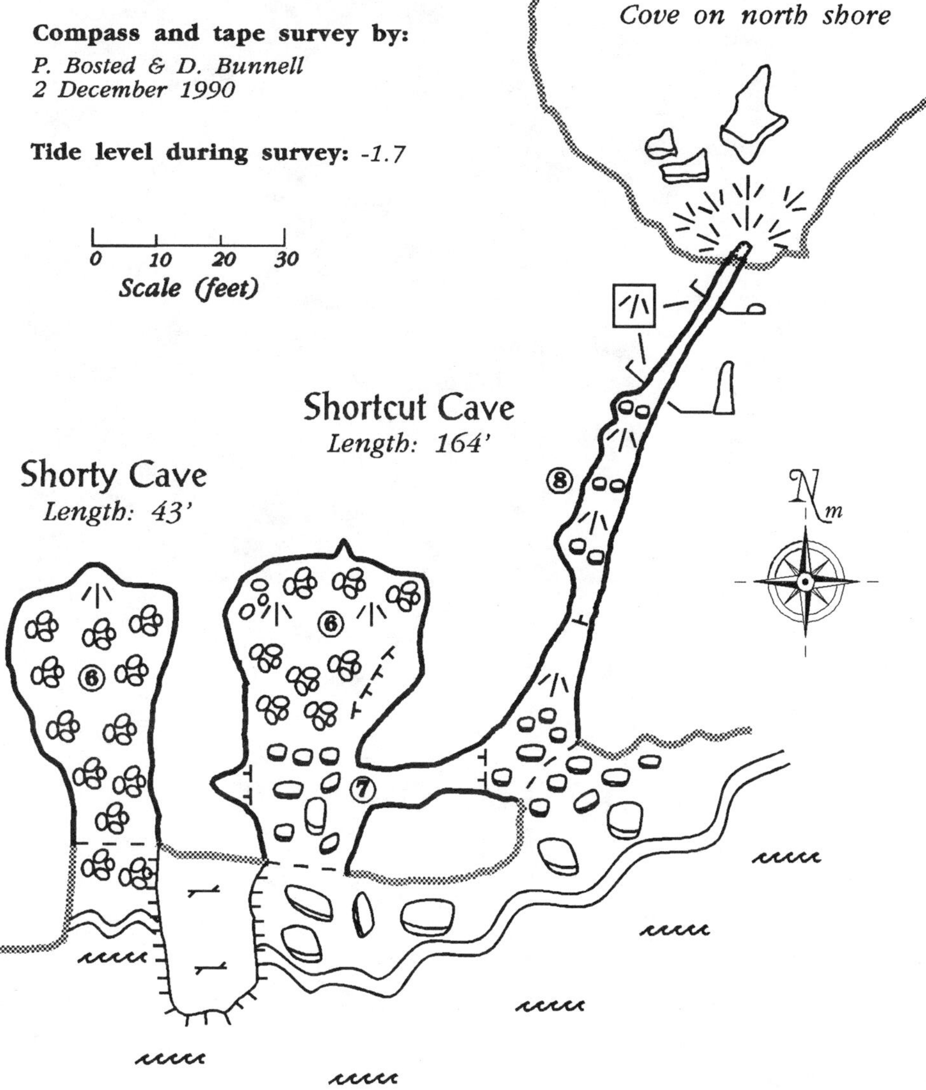
SHORTCUT & SHORTY CAVES
West Anacapa Island, California
Channel Islands National Park
Compass and tape survey by:
P. Bosted & D. Bunnell
2 December 1990
Tide level during survey: -1.7
0 10 20 30
Scale (feet)
Cove on north shore
Shortcut Cave
Length: 164'
Shorty Cave
Length: 43'
Produced by Dave Bunnell

The author crawls out of the north entrance of Shortcut Cave

CAT'S SHADOW CAVE - 35

Location: On the south shore just across from Cat's Rock

Entrances: 1

Length: 86'

Conditions: The cave can be reached by hiking from Frenchy's cove at a very low tide, at which time the cave is largely emptied out. However, entry may still involve getting wet. Not suitable for exploration by dinghy or kayak, due to rocks in the entrance zone. A light would only be needed in the far reaches of the cave on the cobble beach.

Description: The 15' high x 15' wide entrance leads into a pretty little cave. It is all walking-height and has a nice sculpted cross-section, probably owing to the presence of large cobbles in the rear that have helped to scour it out.

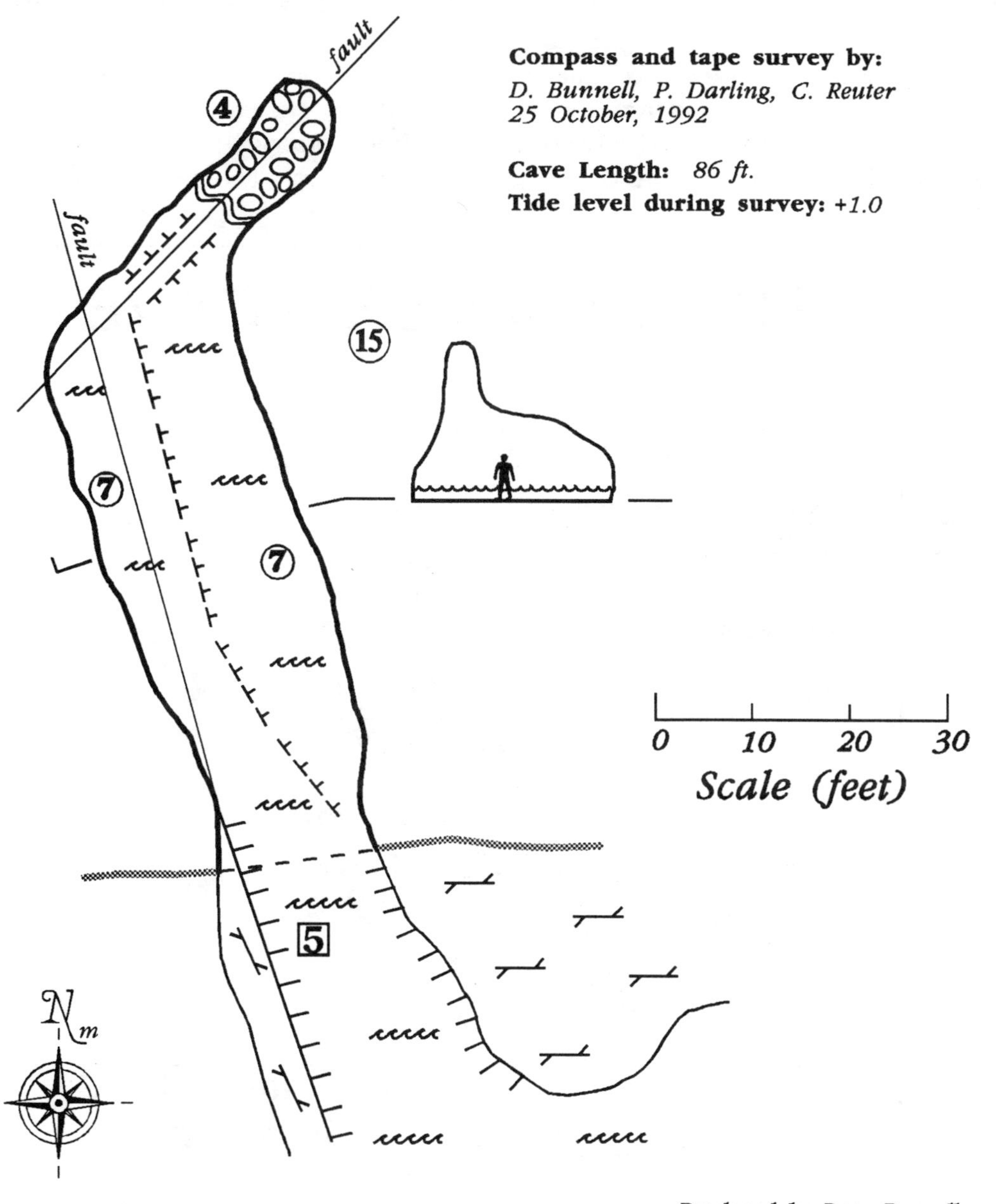
CAT'S SHADOW CAVE
West Anacapa Island, California
Channel Islands National Park
Compass and tape survey by:
D. Bunnell, P. Darling, C. Reuter
25 October, 1992
Cave Length: 86 ft.
Tide level during survey: +1.0
fault
fault
4
15
7
7
5
0 10 20 30
Scale (feet)
N
m
Produced by Dave Bunnell

LITTLE SUNBEAM CAVE - 36

Location: On the south shore about 800' west of Cat Rock

Entrances: 2

Length: 86'

Conditions: The cave can be reached by landing a kayak or dinghy on the rock nearby benches at a low tide, at which time the cave is largely emptied out. However, entry may still involve getting wet. A light would be needed to explore the two small branches of the cave .

Description: The cave can be entered through the surge channel leading into it or by climbing down a small hole from the rock bench above. The cave's main room is some 15' high, with a higher domed ceiling on the left. Some orange flowstone cascades down the left wall. Beyond the main room, the cave branches into two narrow cobble-floored fissures.

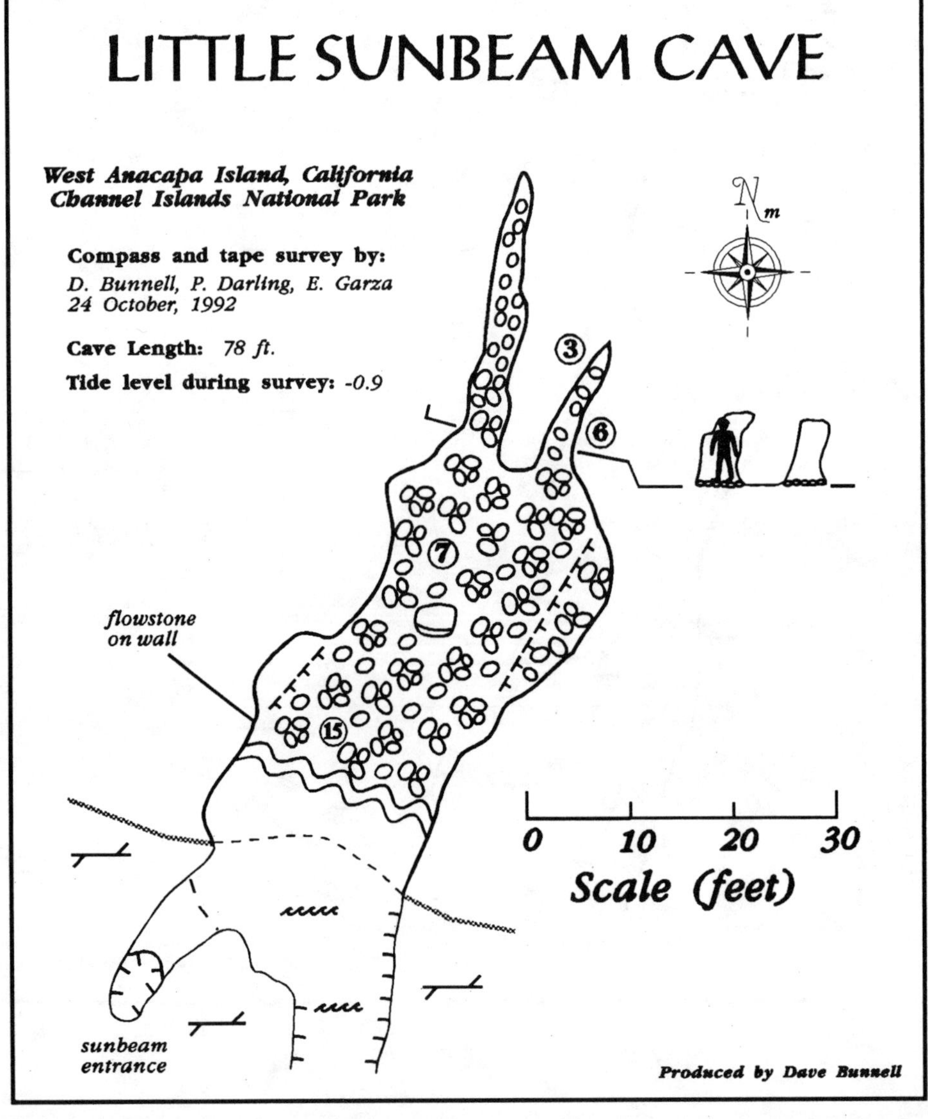

PHIL'S FOLLY CAVE - 37

Location: On the south shore about 900' west of Cat Rock

Entrances: 1

Length: 110'

Conditions: The cave can be reached by landing a kayak or dinghy on the rock nearby benches at a low tide, which is the only feasible time to explore this cave. The lower the tide, the better for this cave. Don't visit if any appreciable swell is running as the surge channel leading into the cave tends to amplify the swell. A light is definitely useful in this dark, narrow cave.

Description: At high tides, this cave is submerged, becoming a blowhole as the tide lowers and then partially emptying out. The entrance is at the back of a wide surge channel which contains a large boulder. The 6' high entrance and the 7' high passage beyond are completely coated with red algae. The passage extends as a straight shot to a cobble beach, which was still in the surf zone during our visit.

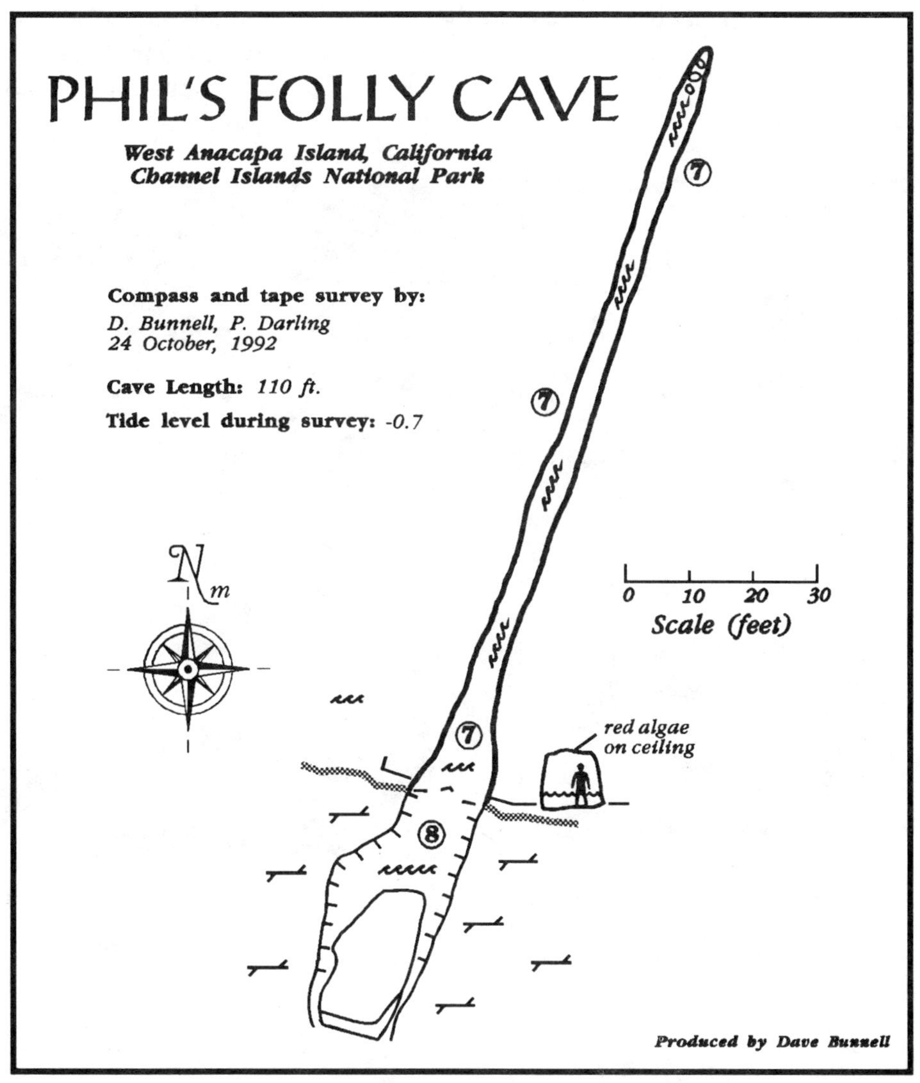

CAT'S EYE CAVES #1-7 -(38-44)

Location: On the south shore of West Island, about 1,000' west of Cat Rock

Entrances: 1 in each of the 7 caves

Lengths: 58', 62', 75', 34', 40', 65', 58'

Conditions: All of these caves are best entered at low tide and when seas are calm, as most have surge channels in front. We walked between the entrances on the bedrock benches but had to wade or swim the surge channels leading into the caves. Lights are useful in #4, #5, and #6.

Descriptions: These caves are all in close proximity to each other yet none is very large, so they are depicted on a single map. This is at the expense of detail but it demonstrates how the forming faults parallel one another. All of the caves are basic single passages with no branches or side tunnels. All of them had either a sand or cobble beach in the rear at low tide. Cave #4 is somewhat different, with a pillar dividing the passage. Cave #6 has a very narrow passage which one can either squeeze into at the front, or else chimney along high and drop into the back. It had a nice rounded bedrock tunnel curving off the main fissure in its rear, ending in the large cobbles usually associated with such smooth tunnels. Some of the caves, especially #6 and #7, were home to large colonies of honeycomb worms. The entrance cross-section of #7 shows an unusual rounded ceiling, as seen in the photo.

Cat's Eye Caves #5, #6, and #7. Note the person standing just to the right of the entrance to #7

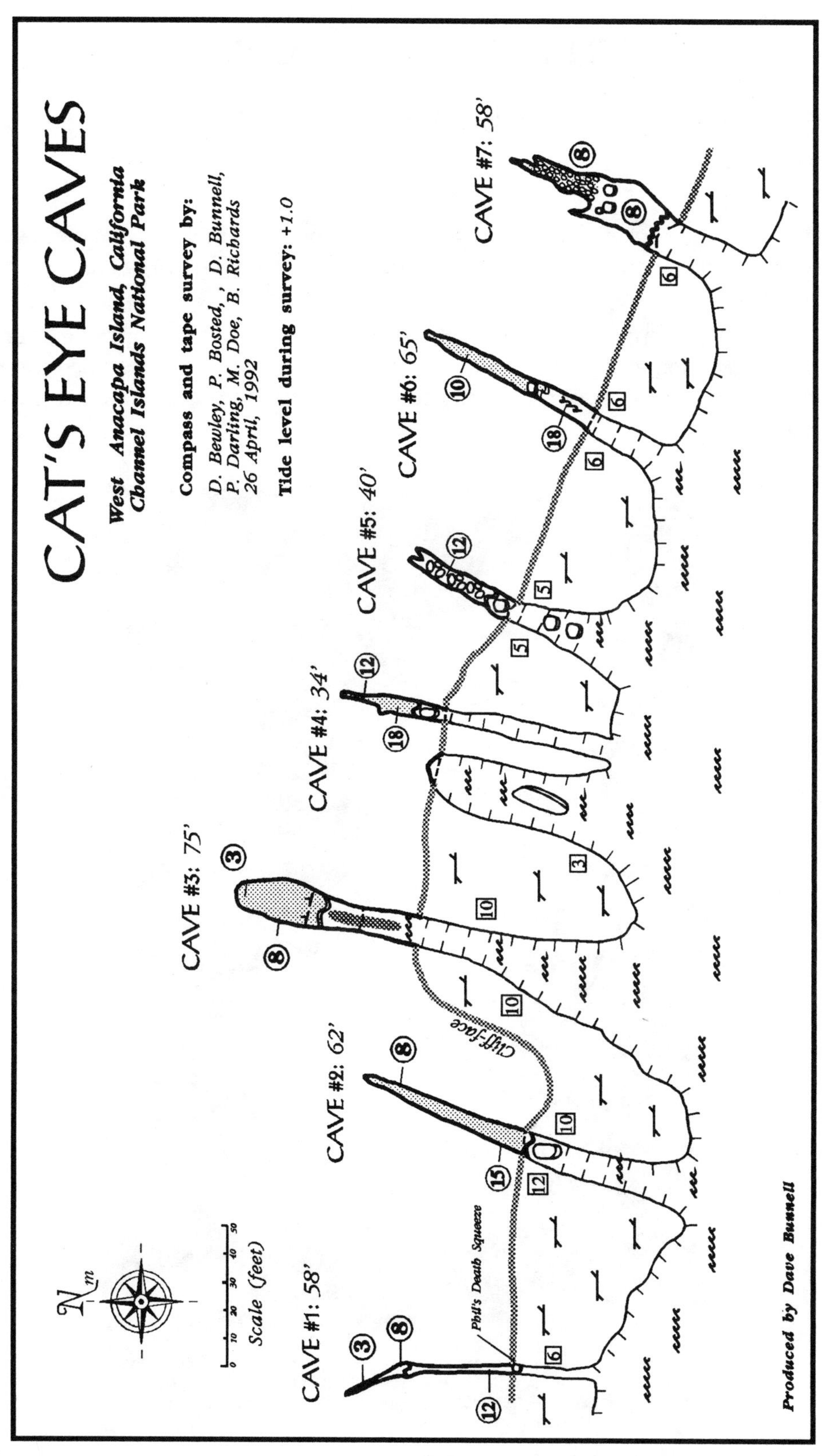
CAT'S EYE CAVES
West Anacapa Island, California
Channel Islands National Park
Compass and tape survey by:
D. Bewley, P. Bosted, , D. Bunnell, P. Darling, M. Doe, B. Richards
26 April, 1992
Tide level during survey: +1.0
CAVE #1: 58'
CAVE #2: 62'
CAVE #3: 75'
CAVE #4: 34'
CAVE #5: 40'
CAVE #6: 65'
CAVE #7: 58'
Phil's Death Squeeze
Cliff-face
Scale (feet)
0 10 20 30 40 50
Produced by Dave Bunnell

LONELY AT THE TOP CAVE - 45

Location: 2000' from the tip of West Island.

Entrances: 1

Length: 116'

Conditions: The cave is dry but requires a swim to reach. In very calm conditions one might pull a kayak up on the rock shelves in front of the cave. The large entrance provides sufficient light to explore it.

Description: The entrance is large (28' wide and 40' high) and leads one over a floor of broken rock fragments for 70' to a very steeply sloping sand pile. The cave ends at the top of this pile, whose source is a mystery. However, one does get a nice view of the cave from the top.

View from the sand slope in the rear of Lonely at the Top

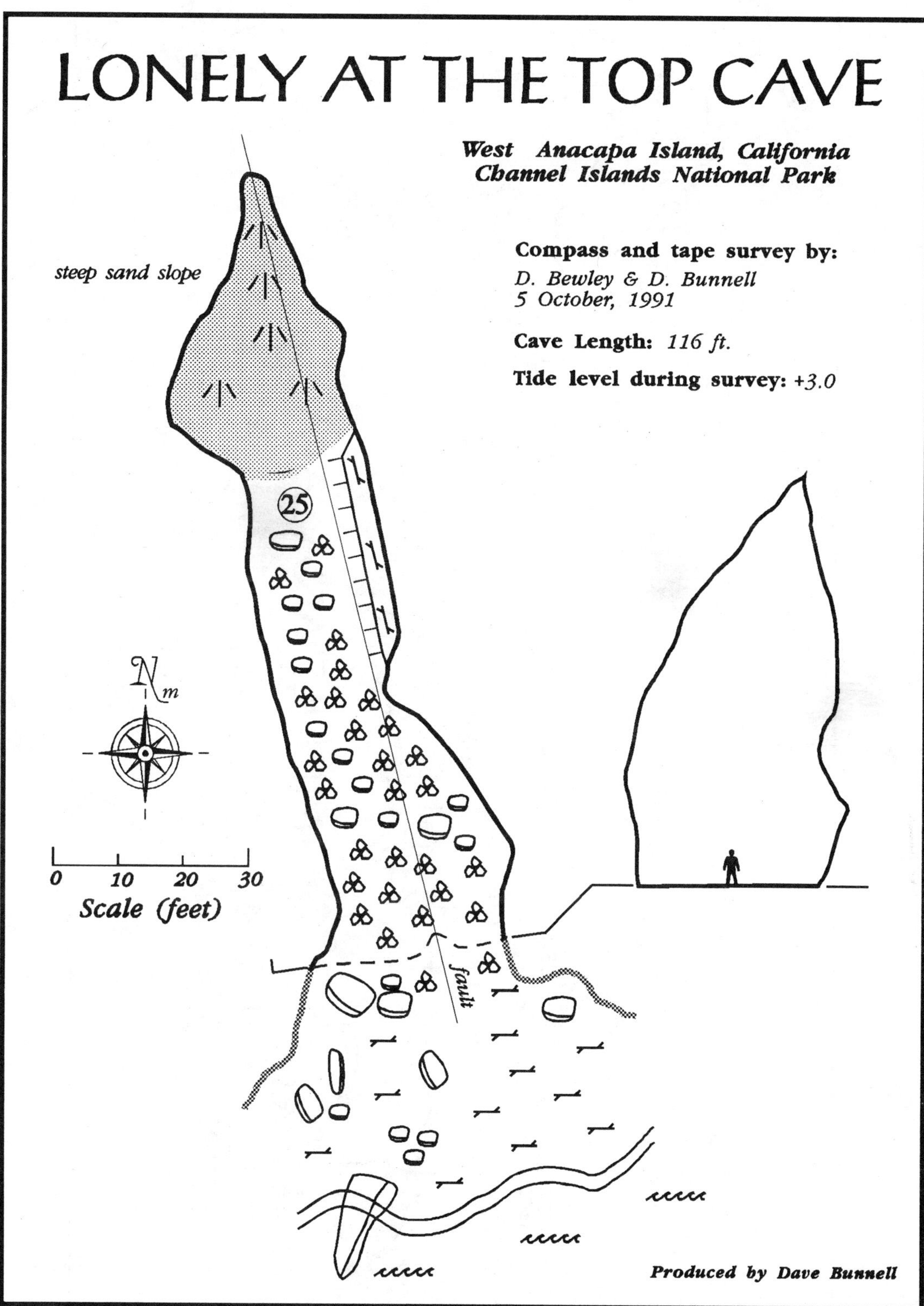
LONELY AT THE TOP CAVE
West Anacapa Island, California
Channel Islands National Park
Compass and tape survey by:
D. Bewley & D. Bunnell
5 October, 1991
Cave Length: 116 ft.
Tide level during survey: +3.0
steep sand slope
25
fault
N m
0 10 20 30
Scale (feet)
Produced by Dave Bunnell

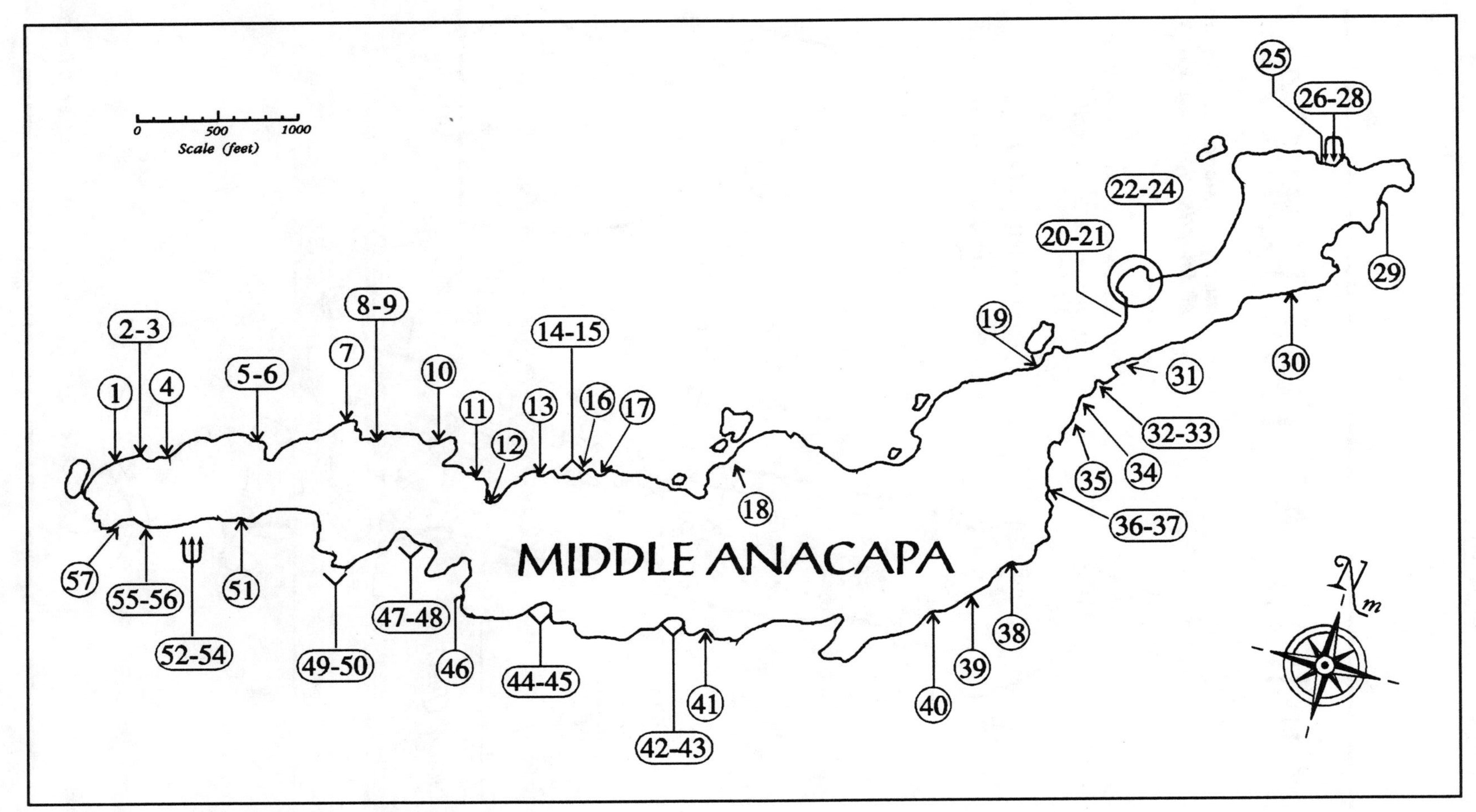
0 500 1000
Scale (feet)
MIDDLE ANACAPA
1
2-3
4
5-6
7
8-9
10
11
12
13
14-15
16
17
18
19
20-21
22-24
25
26-28
29
30
31
32-33
34
35
36-37
38
39
40
41
42-43
44-45
46
47-48
49-50
51
52-54
55-56
57
N m

MIDDLE ISLAND CAVES

NORTH SHORE

1 DISAPPEARING BEACH CAVE (63')
1a {cave feature, left-angle fault}
2 KELPFLY CAVE (116')
3 ATTIC CAVE (206')
4 THE AERIE (164')
4a {cave feature, chasm w/white above}
5 UPTIGHT CAVE (37')
6 DOWNVIEW CAVE (142')
7 HIDDEN TUNNEL (187')
8 TARRY SHELVES CAVE (130')
9 DANGLING KELP CAVE (89')
10 KEYHOLE ROCK CAVE (590')
11 UPSTAIRS-DOWNSTAIRS CAVE (83')
12 IRON CHOCKSTONE CAVE (232')
13 HIDDEN PASSAGE CAVE (130')
14 GURGLING CAVE (90.5')
15 SELDOM SEEN CAVE (114')
16 BATSTAR CAVE (97')
17 AGATE BEACH (SIGN) CAVE (136')
18 COMPLEX CHASM (390')
19 CONVERGING FAULTS CAVE (133')
20 LITTLE SURPRISE CAVE (145')
21 EL CHIQUITO (34')
22 WESTVIEW CAVE (82')
23 TREASURE CHEST CAVE (207')
24 VICTORY CAVE (175')
25 CRAWLSPOT (195')
26 SLOPING COBBLE (56')
27 SAND CAVE (94')
28 UNDERCUT CAVE (119')

SOUTH SHORE

29 SEE-THE-STACK CAVE (73')
30 MORNING THUNDER CAVE (130')
31 REFUGE CAVE (54')
31a {cave feature, 2 short levels}
32 LADIES' DELIGHT CAVE (72')
33 HONEYCOMB WORM CAVE (114')
34 SHIPWRECK CAVE (81')
35 GREEN ABALONE CAVES (33' & 27')
35a {cave feature - wide shelter}
36 FISH CAMP CAVE #1 (74')
36a {cave feature - 12' deep}
36b {cave feature - 21' deep}
37 FISH CAMP CAVE #2 (55')
38 LAVA BENCH CAVE #1 (58')
39 LAVA BENCH CAVE #2 (115')
40 RESPIRING CHIMNEY CAVE (448')
41 SUNBEAM CAVE (172.5')
42 CRABBY CAVE (149')
43 THREE DOOR CAVE (589')
43a {cave feature-high fissure}
44 DOGLEG CAVE (132')
45 SEAL'S BEACH CAVE (84')
46 DEEP PENETRATION (185')
47 COBBLE COVE CAVE #1 (50')
48 COBBLE COVE CAVE #2 (36')
49 TIDAL LAGOON CAVE (58')
50 CLIFF CHASM (36')
51 SANDY SURPRISE (155')
52 URCHINS' LAIR (92')
53 PUD CAVE (55')
54 PNEUMATIC CAVE (124')
55 RIPPLING REFLECTIONS CAVE (117')
56 PURPLE TURMOIL CAVE (85')
57 SLIPPERY ROCK CAVE (73')

DISAPPEARING BEACH CAVE - 1

Location: On the west tip of Middle Anacapa, north shore. A beach appears here at lower tides.

Entrances: 1

Length: 63'

Conditions: At an extreme low tide, the cave is dry. No lights are necessary. At higher tide levels the beach disappears and the cave is floored with water for its entire length. Not suitalbe for exploration by kayak or dinghy.

Description: The entance is 18' wide and 10' high and leads into a sand-floored passage which is mostly walking-height throughout.

Approaching Disappearing Beach Cave at a higher tide, when no beach is present

DISAPPEARING BEACH CAVE

Middle Anacapa Island, California
Channel Islands National Park

Compass and tape survey by:
D. Bunnell, E. Garza, M. Oliphant, N. Pistole
14 October 1989

Cave Length: *63 ft.*

Tide level during survey: *-0.6*

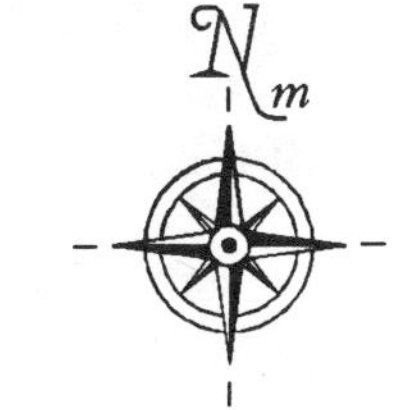

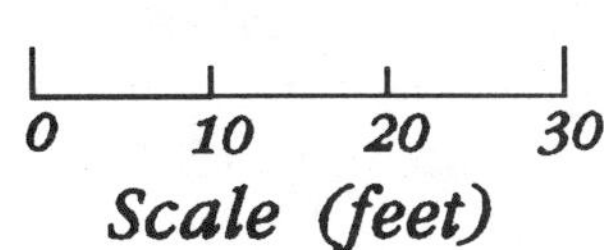

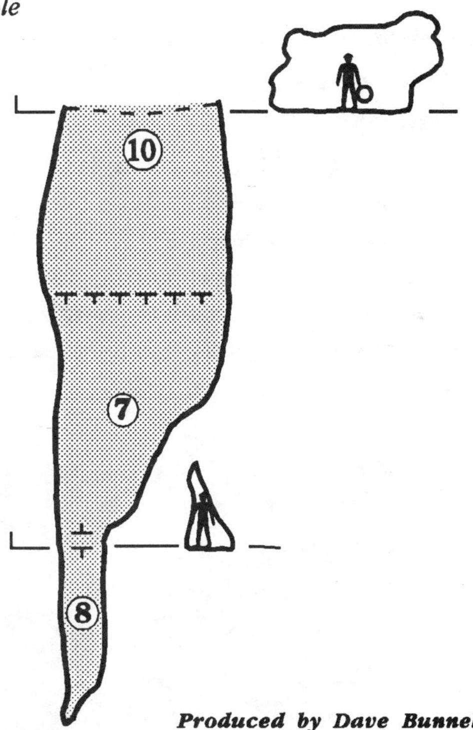

Produced by Dave Bunnell

KELPFLY CAVE - 2

Location: In the westernmost small cove on the north shore of Middle Island.

Entrances: 1

Length: 116.5'

Conditions: Largely dry at low tide. Not readily explorable by dinghy exept for the first 50' or so, but beware submerged rocks. A light is useful in the rear portions.

Description: A typical single-passage fissure cave formed along a fault oriented at 210°. The ceilings are high enough to permit walking but are low in a couple spots, requiring stooping or crawling. The cave was named for the numerous flies congregating on kelp deposited in the cave.

ATTIC CAVE - 3

Location: In the westernmost small cove on the north shore of Middle Island

Entrances: 1

Length: 138'

Conditions: Much of the cave is probably dry even at high tide, and must be explored on foot after reaching the entrance. A light is useful in the rear portions.

Description: This is a single-passage fissure like its neighbor, Kelpfly. However, it differs in having an upper level (the Attic) which is wider than the lower level. Getting up involves a moderately difficult climb on crumbly rock, so caution is advised. The first 40-50' of the cave has collapsed as is evident by the rockfall and the recession of the dripline down the passage.

Kelpfly (on right) and Attic (on left) cave entrances

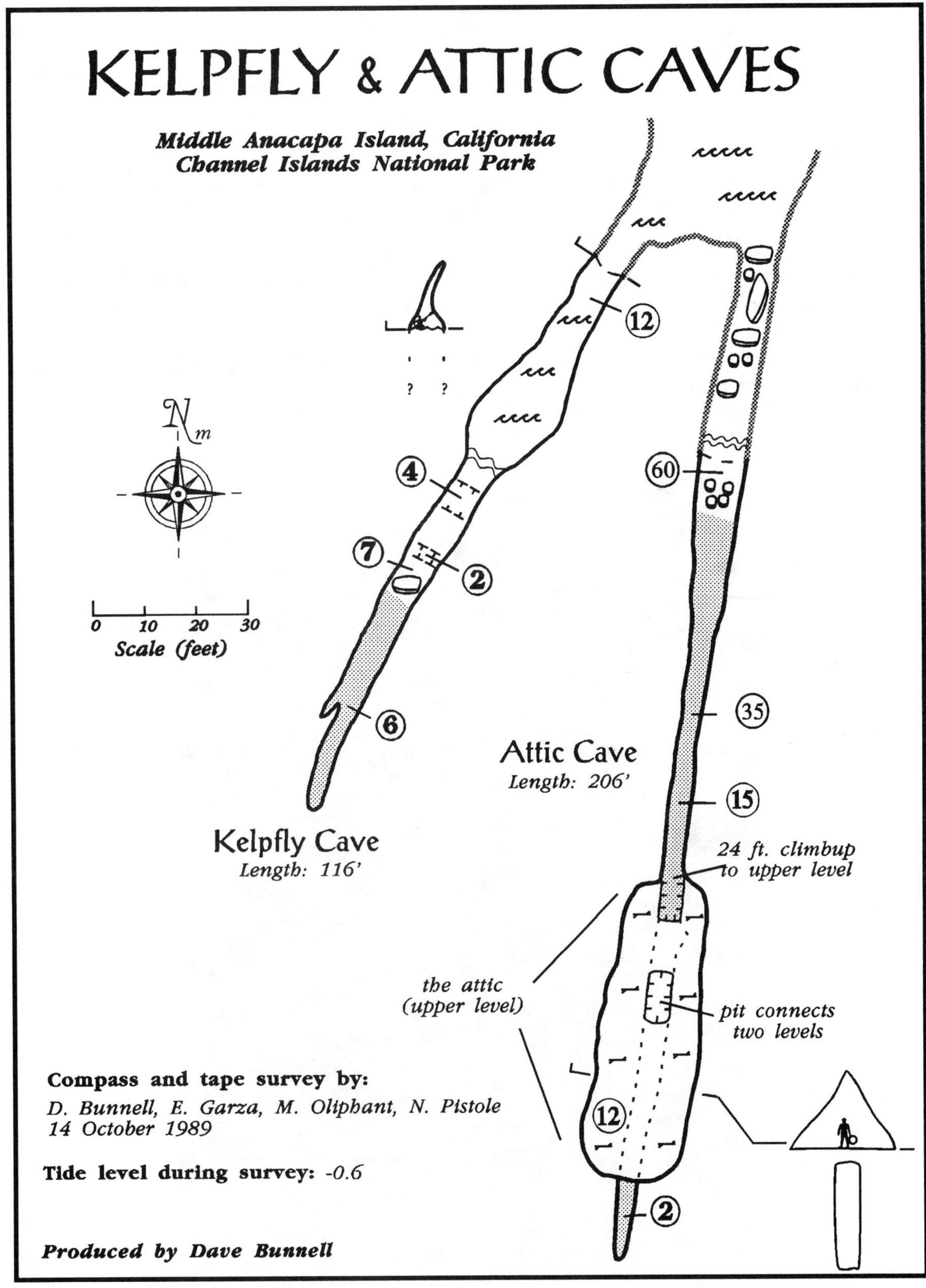
KELPFLY & ATTIC CAVES
Middle Anacapa Island, California
Channel Islands National Park
N m
0 10 20 30
Scale (feet)
12
4
7
2
6
60
35
15
2
12
? ?
Attic Cave
Length: 206'
Kelpfly Cave
Length: 116'
24 ft. climbup to upper level
the attic (upper level)
pit connects two levels
Compass and tape survey by:
D. Bunnell, E. Garza, M. Oliphant, N. Pistole
14 October 1989
Tide level during survey: -0.6
Produced by Dave Bunnell

THE AERIE - 4

Location: 400' east of the western tip of Middle Island, on the north shore

Entrances: 1

Length: 164'

Conditions: The upper portion could be visited at any tide level, whereas the lower portions are largely submerged except at low tide. Lights would probably be needed in the lower level, which we did not explore.

Description: This unusual cave has a large oval entrance some 50' wide and 60' high. The upper level is floored with bedrock and powdery residue and shows considerable evidence of use by birds, as indicated by the accumulated feathers. A surge channel at the entrance has split into two diverging short passages which are actively enlarging beneath the upper level.

Aerie Cave entrance

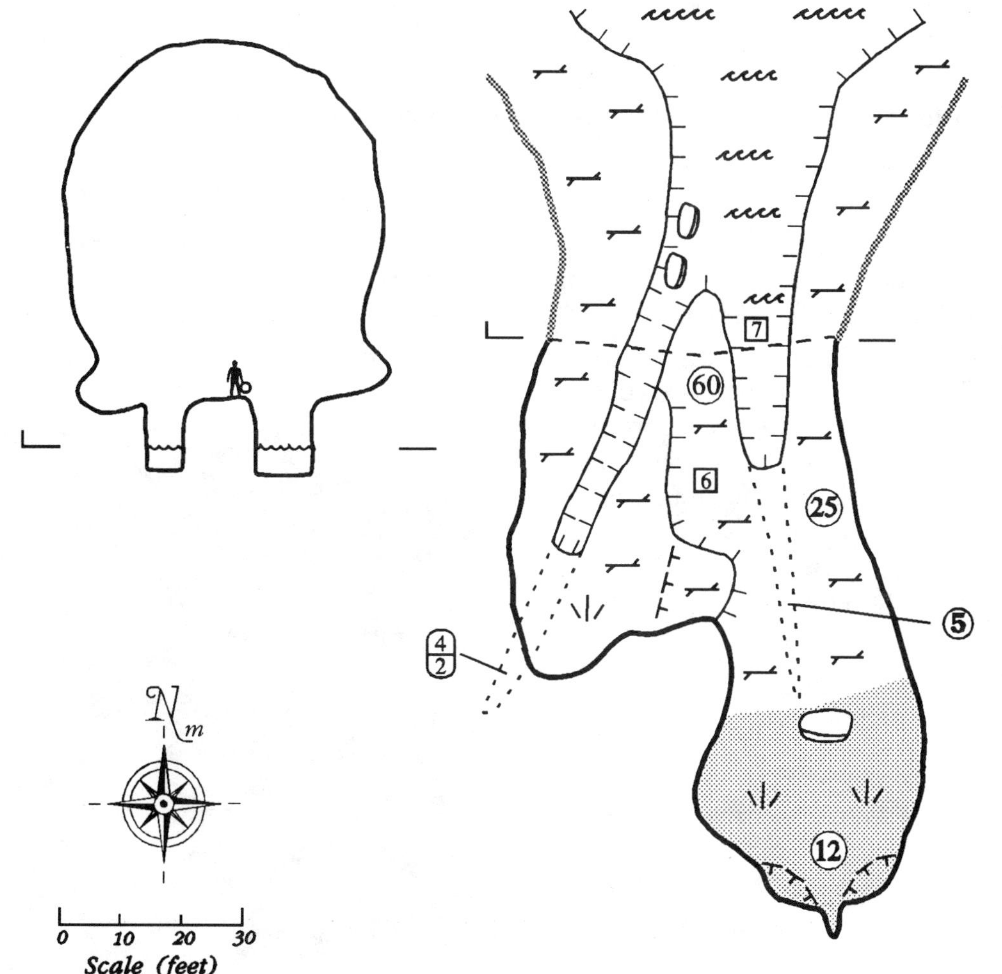
AERIE CAVE
Middle Anacapa Island, California
Channel Islands National Park
Compass and tape survey by:
D. Bunnell, E. Garza, M. Oliphant, N. Pistole
14 October 1989
Cave Length: 164 ft.
Tide level during survey: -0.6
7
60
6
25
5
4
2
12
N m
0 10 20 30
Scale (feet)
Produced by Dave Bunnell

UPTIGHT CAVE - 5

Location: about 800' east of the west tip of Middle Island, on the north shore

Entrances: 1

Length: 37'

Conditions: This cave must be explored at low tide, and must be explored on foot after reaching the entrance. A light is useful in the rear portions.

Description: The dripline has receded down the channel leading into the cave, and collapse from the former roof now partially blocks the entrance. The cave is walking height for 30' to where one can squeeze up into a crevice, beyond which one can see another 15' or so of unpassably tight passage.

DOWNVIEW CAVE - 6

Location: about 800' east of the west tip of Middle Island, on the north shore

Entrances: 2

Length: 142'

Conditions: This cave can be explored at moderate tide levels, with an entrance by kayak or dighy possible if the seas are calm. Otherwise, one must swim or snorkle in. Check conditions inside by observing from the top. A light is useful in the rear portions.

Description: This unusual cave was named for the presence of a skylight which allows one to look down the main axis of the cave. From this vantage point one can see almost 100' into the cave. The cave is water-floored for much of its length but has a nice sandy beach in the rear and passage dimensions are comfortably large.

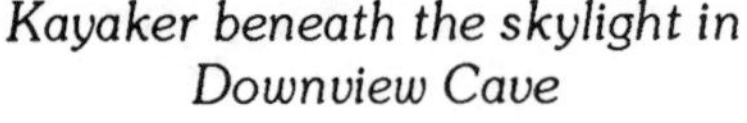

Kayaker beneath the skylight in Downview Cave

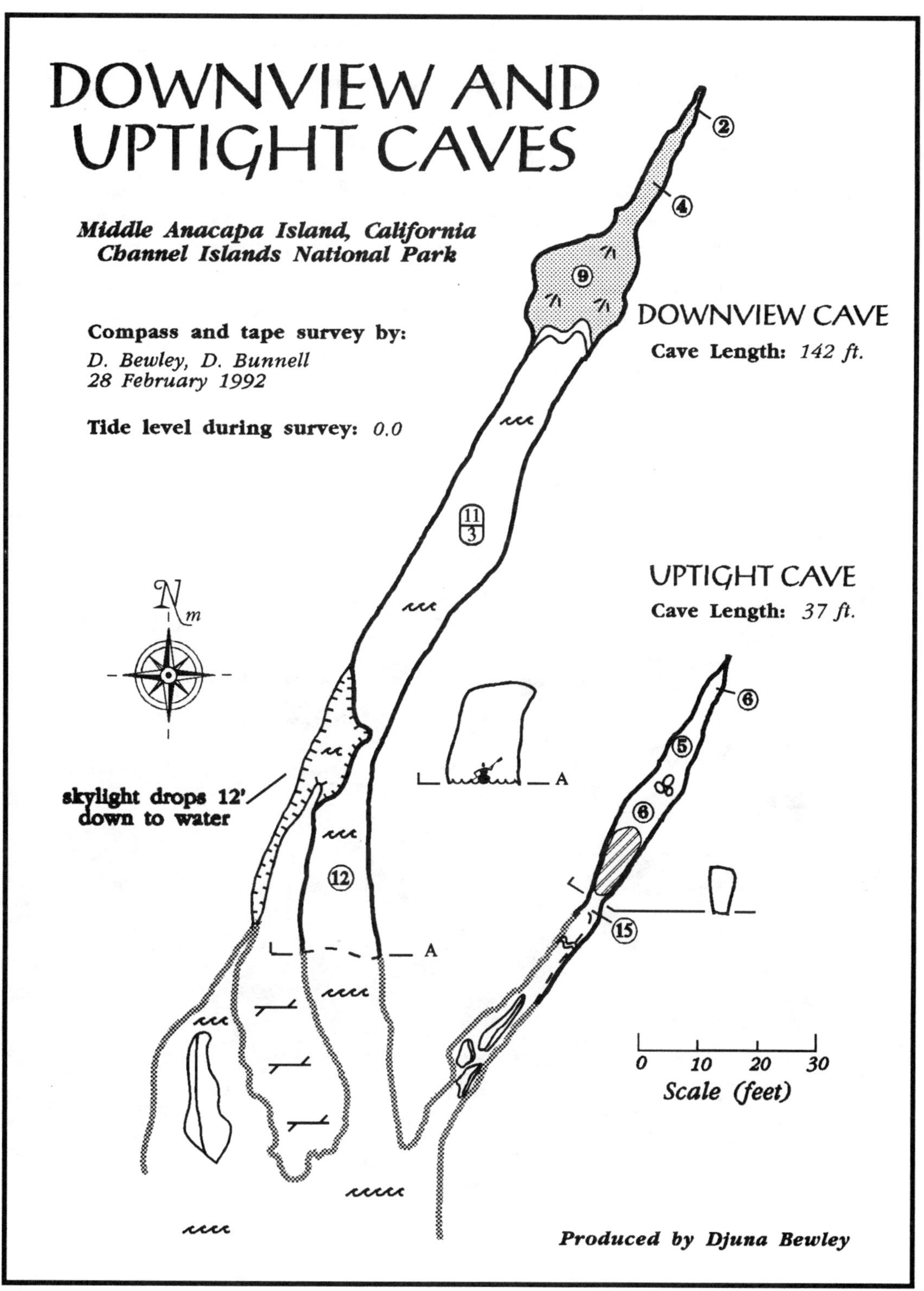
DOWNVIEW AND UPTIGHT CAVES
Middle Anacapa Island, California
Channel Islands National Park
Compass and tape survey by:
D. Bewley, D. Bunnell
28 February 1992
Tide level during survey: 0.0
DOWNVIEW CAVE
Cave Length: 142 ft.
UPTIGHT CAVE
Cave Length: 37 ft.
skylight drops 12' down to water
A
A
0 10 20 30
Scale (feet)
Produced by Djuna Bewley

HIDDEN TUNNEL CAVE - 7

Location: 600' west of Keyhole Rock, on the north shore

Entrances: 2

Length: 187'

Conditions: Explore at low tide, as the passage profile is hourglass-shaped and thus, higher tides make for narrower passage. The cave is well sheltered from most swell. A good light is useful but not essential. This is a wade and swim-through cave only.

Description: The two entrances to this cave are practically hidden behind fallen rocks. An interesting through trip is afforded by this cave, in a passage varying from 4 to 6' wide, and paralleling the cliff face. A side passage leads to the left just inside the eastern entrance, extending some 60' to a cobble beach. It is open only at low tide.

Kayaker in front of Hidden Tunnel's east entrance

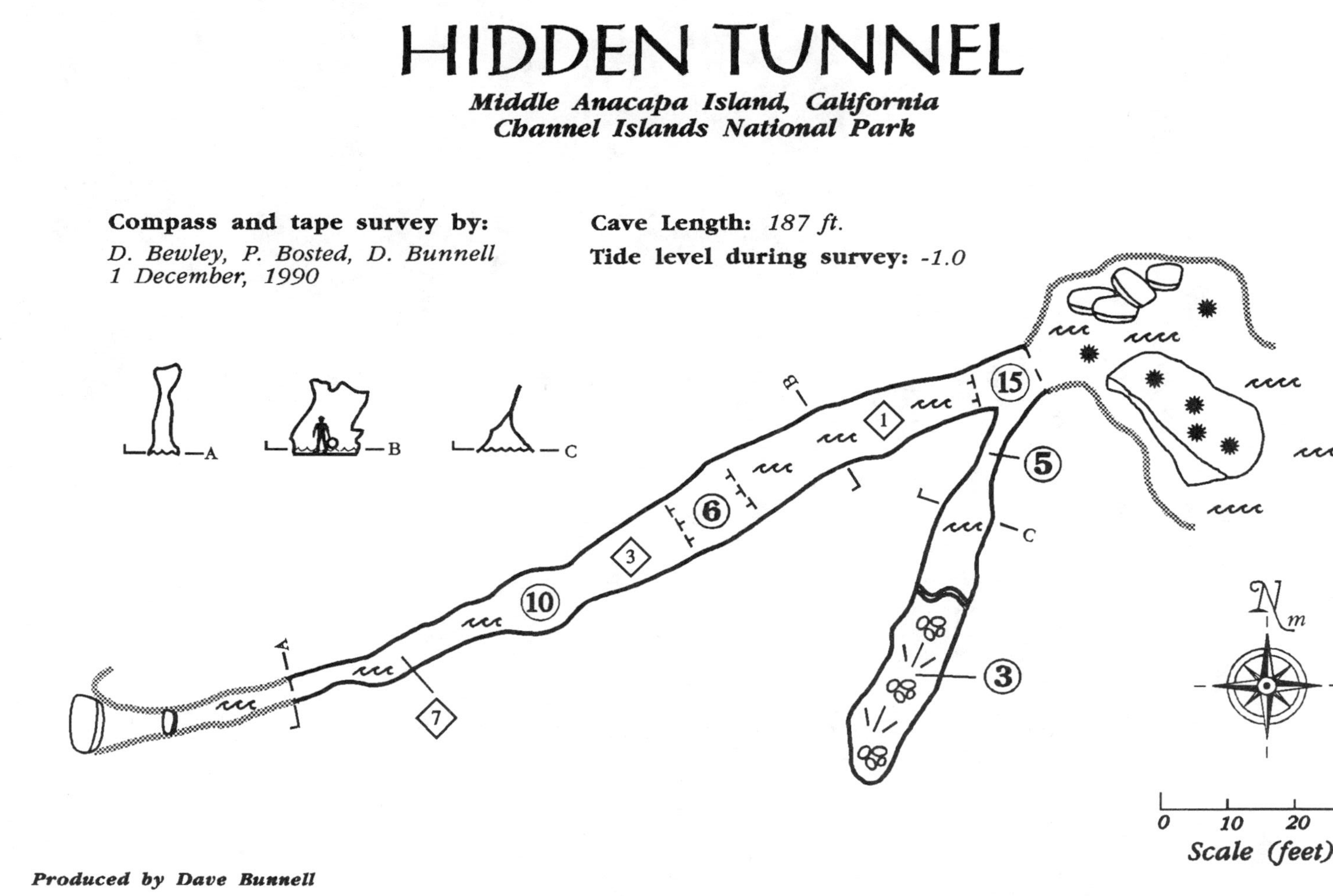
HIDDEN TUNNEL
Middle Anacapa Island, California
Channel Islands National Park
Compass and tape survey by:
D. Bewley, P. Bosted, D. Bunnell
1 December, 1990
Cave Length: 187 ft.
Tide level during survey: -1.0
A
B
C
15
1
5
6
3
10
7
3
m
0
10
20
30
Scale (feet)
Produced by Dave Bunnell

TARRY SHELVES CAVE - 8

Location: 500' west of Keyhole Rock

Entrances: 1

Length: 130'

Conditions: A handline is essential for climbing down into this cave, except at an extremely low tide, when one can swim through a keyhole beneath chockstones. Calm seas are required as the long surge channel amplifies the swells entering the cave. A light is needed for exploring the last 60' or so.

Description: This is a cave which has lost much of its length through ceiling collapse, leaving a long surge channel. One enters from the tar-covered shelves some 10' above water level, by negotiating a slippery climb. The passage beyond the drip line narrows to some 6' wide and is 7 to 8' high for most of its length. The last 60' of the cave is a fissure 6' high and 2' wide, floored with sand.

DANGLING KELP CAVE - 9

Location: 450' west of Keyhole Rock

Entrances: 1

Length: 89'

Conditions: This is a "swim-in" type cave, largely accessible only at low tide. The rear portions warrant a light.

Description: This is a standard single-passage cave on a prominent fault, named for a strand of kelp seen dangling from the ceiling. The entrance is 12' high and 5' wide and leads into a passage which curves to the right. A low spot two-thirds of the way in leads into the small terminal room.

Kayaker in front of Dangling Kelp Cave

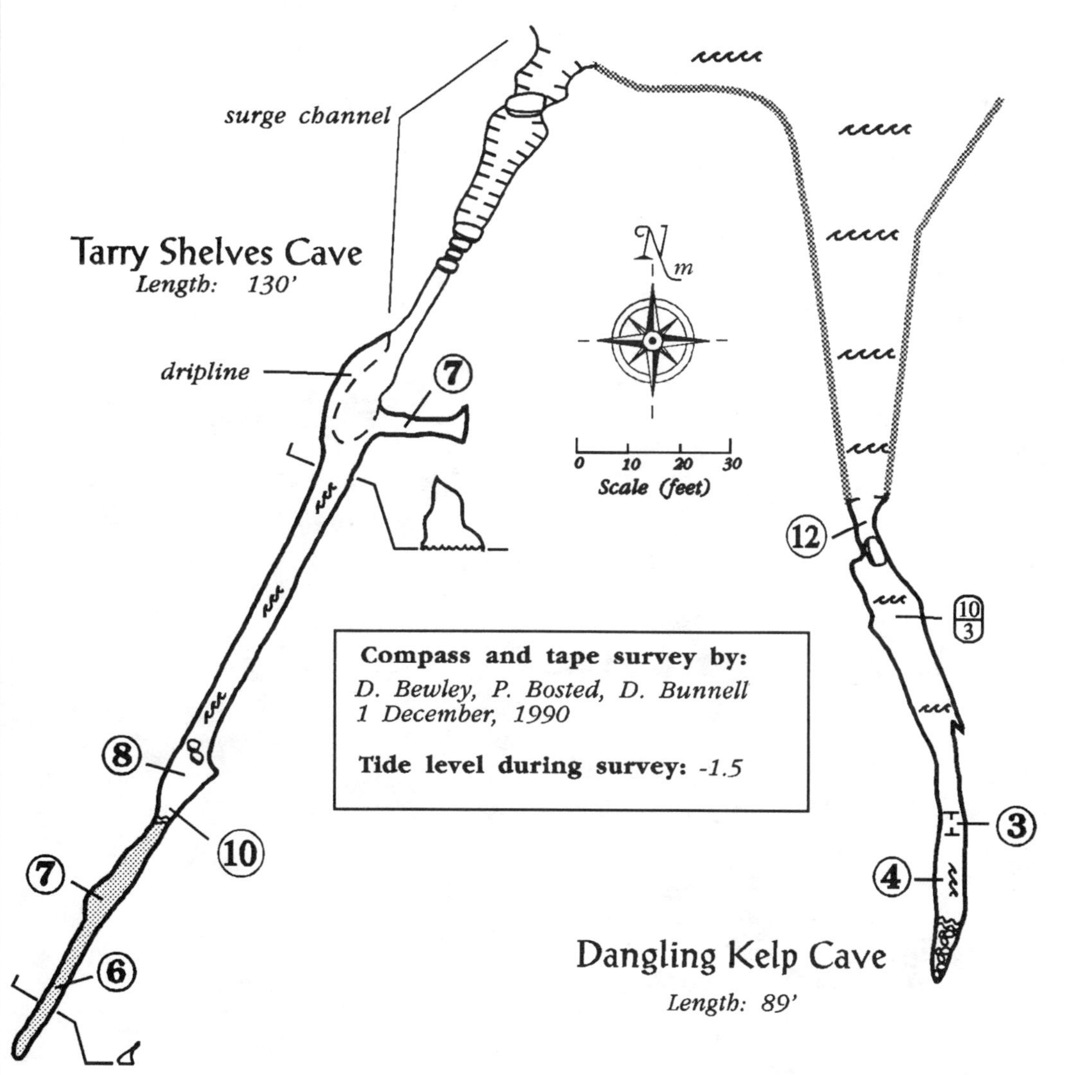
TARRY SHELVES & DANGLING KELP CAVES
Middle Anacapa Island, California
Channel Islands National Park
surge channel
Tarry Shelves Cave
Length: 130'
dripline
7
8
10
7
6
Scale (feet)
0 10 20 30
Compass and tape survey by:
D. Bewley, P. Bosted, D. Bunnell
1 December, 1990
Tide level during survey: -1.5
12
10
3
3
4
Dangling Kelp Cave
Length: 89'
Produced by Dave Bunnell

KEYHOLE ROCK CAVE - 10

Location: Just to the right (west) of Keyhole Rock, a 100' high arch marked in the National Park Service's brochure

Entrances: 1

Length: 590'

Conditions: The upper level is dry and can be explored without a light. Care must be taken in traversing over the skylights. The lower level requires a low tide and swimming to explore, along with strong lights and a helmet.

Description: This is an unusual two-level cave. The relict upper level extends some 100' in from the dripline and is 15 to 25' high ending at a steep rock slope in the rear. A long narrow slot in the floor drops 18' into the passage below. The lower passage is more extensive and begins as a 6' wide water-floored fissure averaging 10-15' high. Near the opening to the upper level is a low spot which closes off at high tide and intermittently opens and closes with the swell at low tide. About 100' beyond the low spot, a chamber is reached with a broad cobble beach at the rear. Here a second fault is intersected, creating passages to the left and right which end in sand beaches. A large pillar divides the room further. On the west wall of the room are two small passages which intersect and are traversible only at a very low tide.

Keyhole Rock and Keyhole Rock Cave, on the right

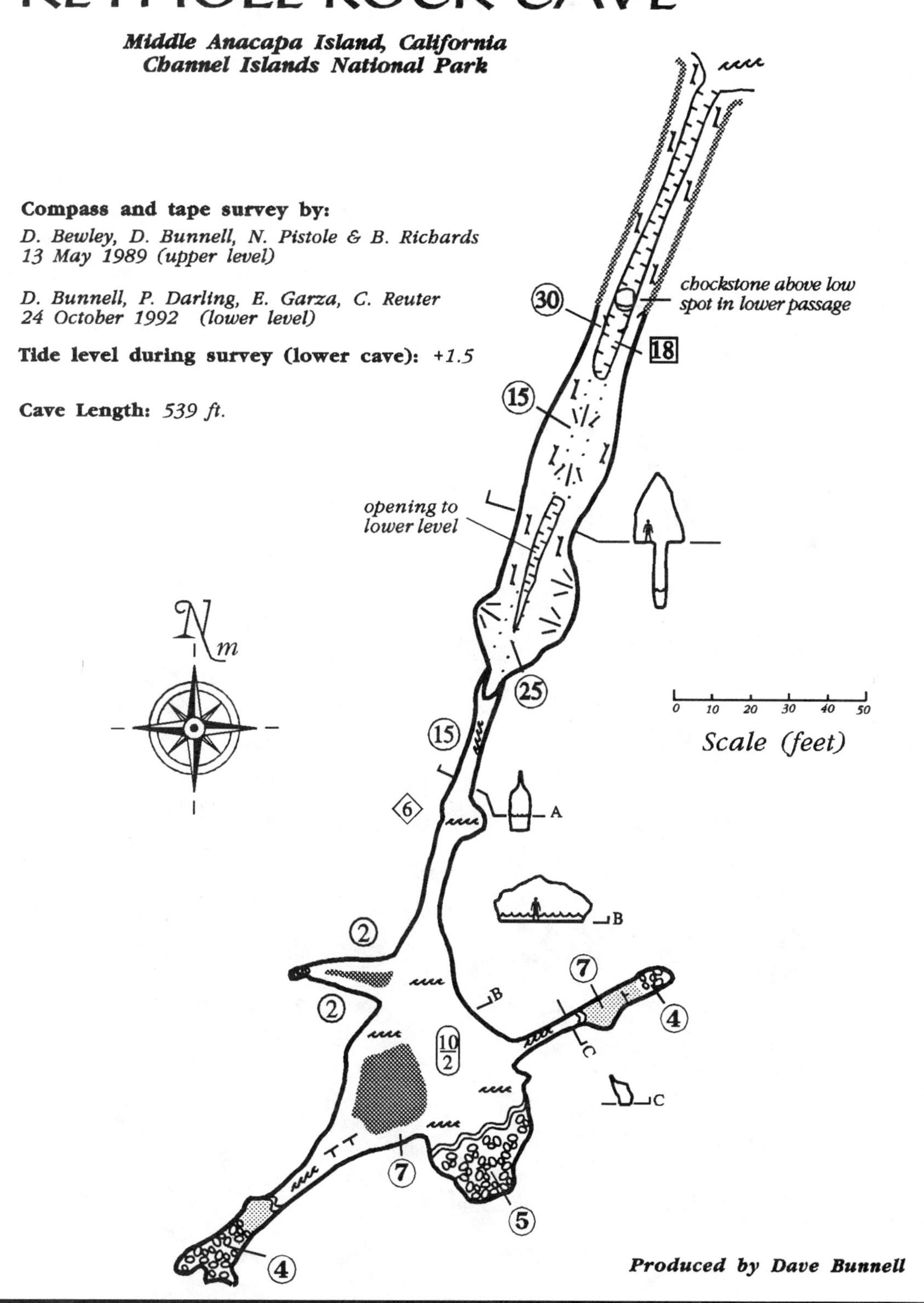
KEYHOLE ROCK CAVE
Middle Anacapa Island, California
Channel Islands National Park
Compass and tape survey by:
D. Bewley, D. Bunnell, N. Pistole & B. Richards
13 May 1989 (upper level)
D. Bunnell, P. Darling, E. Garza, C. Reuter
24 October 1992 (lower level)
Tide level during survey (lower cave): +1.5
Cave Length: 539 ft.
chockstone above low spot in lower passage
30
18
15
opening to lower level
25
15
6
A
B
2
2
7
4
C
10/2
7
5
4
0 10 20 30 40 50
Scale (feet)
Produced by Dave Bunnell

UPSTAIRS-DOWNSTAIRS CAVE - 11

Location: About 300' east of Keyhole Rock

Length: 83'

Entrances: 1

Conditions: The upper level is dry and can be explored without lights. The lower level is wet and narrow, and should be explored only in very calm seas at a low tide.

Description: The cave has an upper dry level much used by birds. It extends back about 70', with ceiling heights up to 15'. In the rear it narrows to a 3' high, 2' wide crawl. The lower level is a very narrow water-floored fissure which we followed for about 40'. The passage continues and might be followed further if swell was sufficiently low, or perhaps with SCUBA.

The two levels of Upstairs-Downstairs Cave

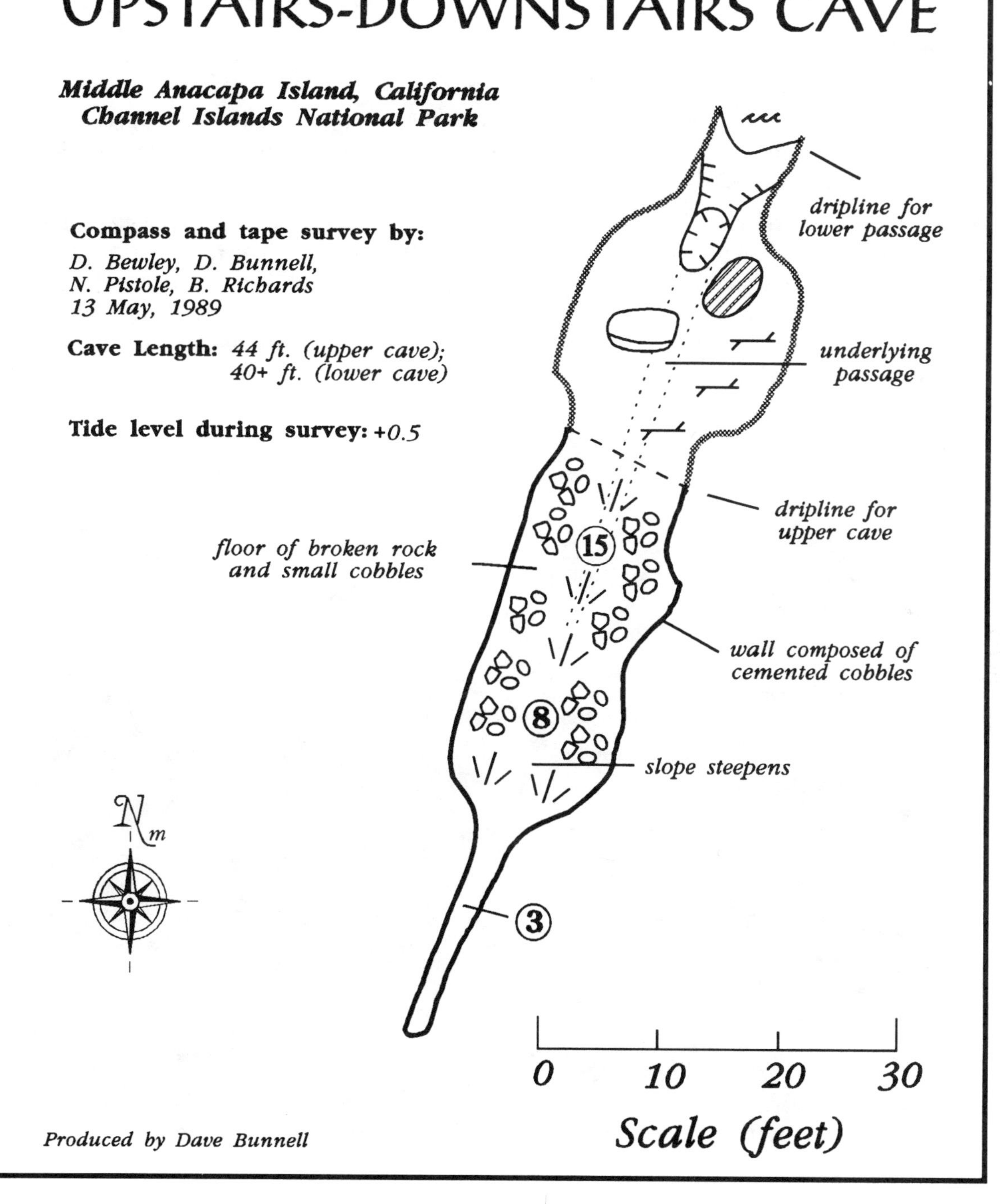
UPSTAIRS-DOWNSTAIRS CAVE
Middle Anacapa Island, California
Channel Islands National Park
Compass and tape survey by:
D. Bewley, D. Bunnell,
N. Pistole, B. Richards
13 May, 1989
Cave Length: 44 ft. (upper cave);
40+ ft. (lower cave)
Tide level during survey: +0.5
dripline for
lower passage
underlying
passage
dripline for
upper cave
floor of broken rock
and small cobbles
15
wall composed of
cemented cobbles
8
slope steepens
3
0 10 20 30
Scale (feet)
Produced by Dave Bunnell

IRON CHOCKSTONE CAVE - 12

Location: In an indented cove 400' east of Keyhole Rock, on the north shore of Middle Island. Also indicated on the USGS topo.

Length: 232'

Entrances: 2

Conditions: The main entrance can be entered in a dinghy or kayak, which could then be beached on small cobble beaches which appear at low tide only. A light is useful to explore the side passages.

Description: The cave has two entrances, the right-hand one being larger and more prominent. Two faults are visible in the cliff above this entrance. The main entrance leads into a water-floored chamber some 50' wide and 15' high. A second passage, 10-12' high, leads into the room from the left-hand entrance, which closes off at high tide. Wedged in the ceiling of this passage at the time of our survey were a spherical iron mooring buoy and a scuba tank.

Both entrances of Iron Chockstone are visible in this photo

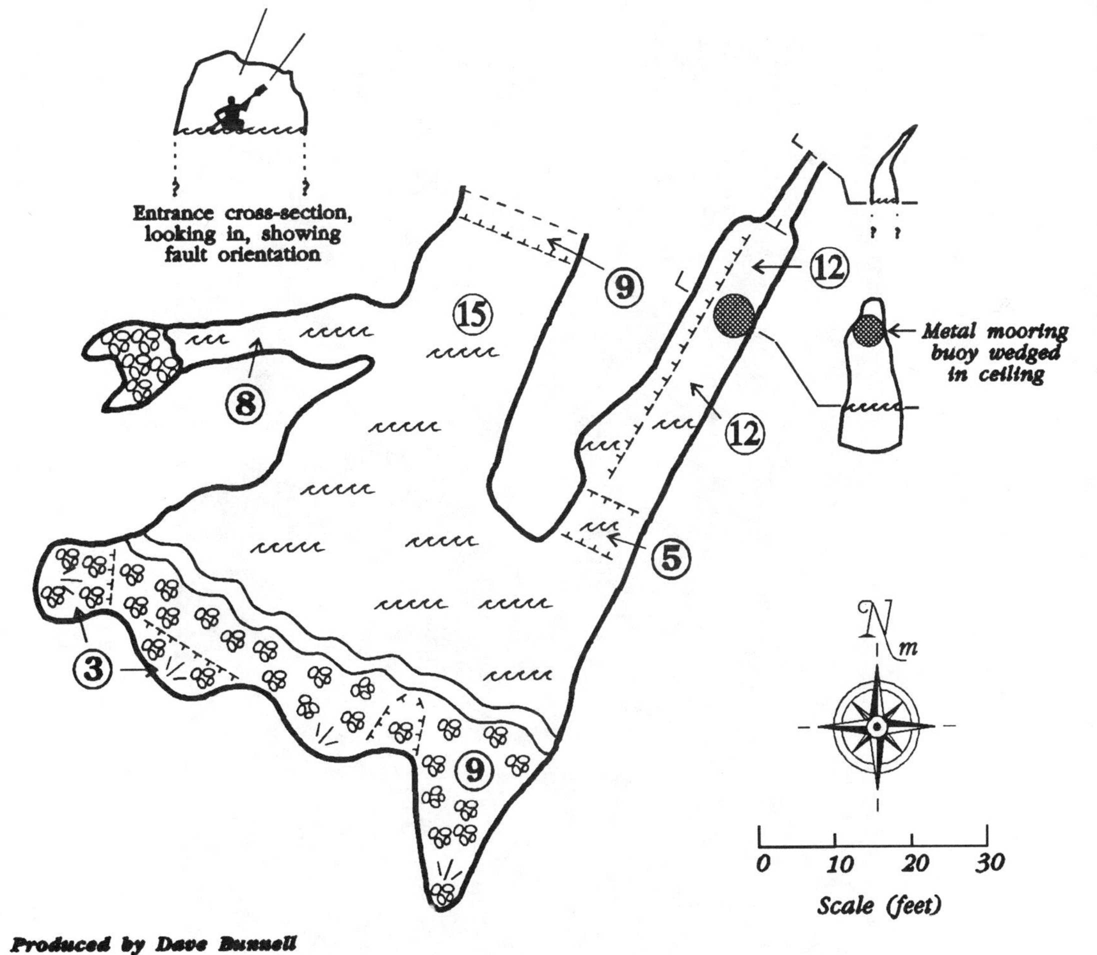
IRON CHOCKSTONE CAVE
Middle Anacapa Island, California
Channel Islands National Park
Compass and tape survey by:
D. Bewley, D. Bunnell, N. Pistole, B. Richards
13 May, 1989
Cave Length: 232 ft.
Tide level during survey: +0.5
Entrance cross-section, looking in, showing fault orientation
15
9
12
12
8
5
3
9
Metal mooring buoy wedged in ceiling
0 10 20 30
Scale (feet)
Produced by Dave Bunnell

HIDDEN PASSAGE CAVE - 13

Location: 700' east of Keyhole Rock

Length: 147'

Entrances: 1

Conditions: The cave is entirely water-floored. A light is necessary to explore the "hidden" passage. The cave might be partially explored in a dinghy but there is nowhere to land it. A low tide and calm seas are necessary to see the cave in its entirety.

Description: The 15' high entrance widens into a small chamber which ends in a narrow fissure which may continue beyond our survey. On the left-hand side of the room a small opening leads into 10' high fissure passage with considerable wind emanating with each surge - probably due to compression of air in the narrow continuation of the passage.

Just inside the entrance; the "hidden passage" is through the small fissure just to the left of the kayak

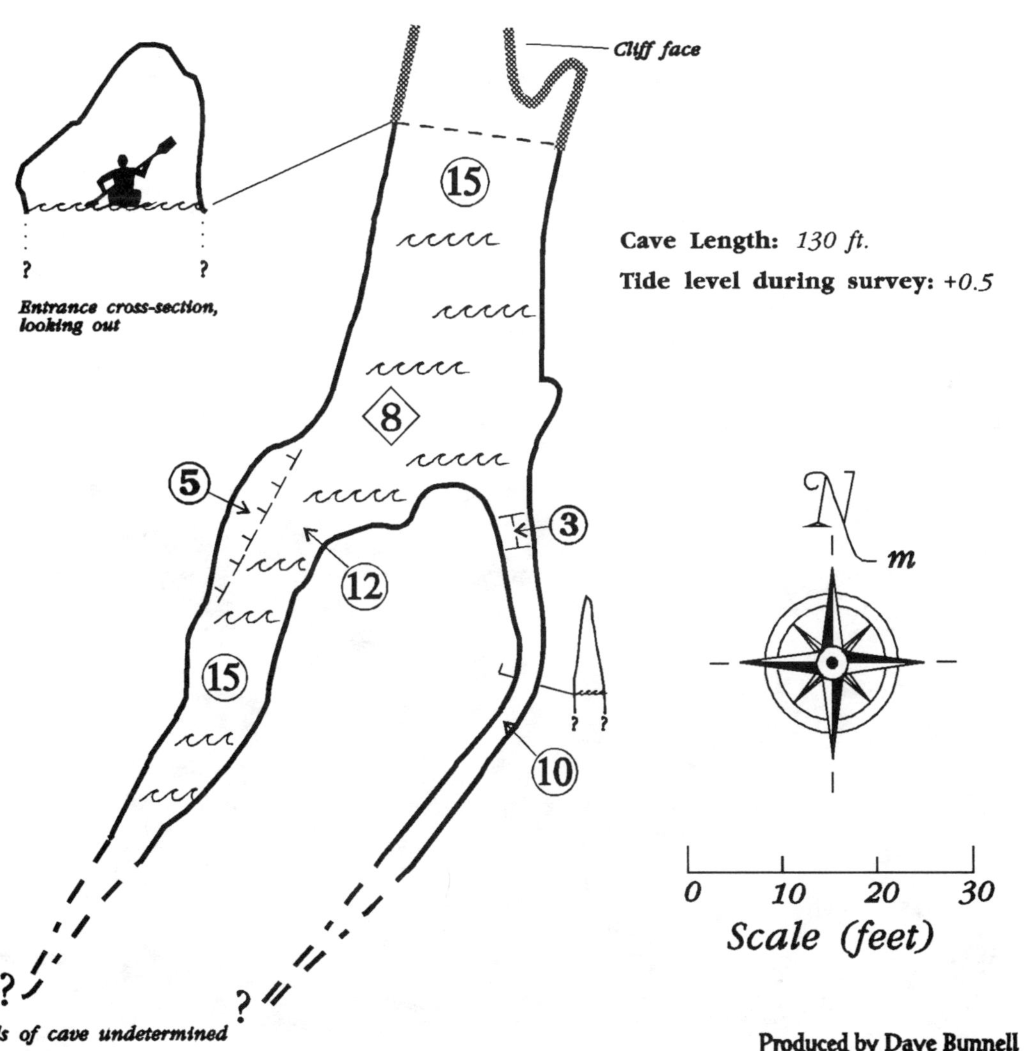
HIDDEN PASSAGE CAVE
Middle Anacapa Island
Channel Islands National Park
Compass and tape survey by:
D. Bewley, D. Bunnell, N. Pistole, B. Richards
13 May, 1989
Cliff face
15
8
5
12
3
15
10
Entrance cross-section, looking out
Cave Length: 130 ft.
Tide level during survey: +0.5
m
0 10 20 30
Scale (feet)
Ends of cave undetermined
Produced by Dave Bunnell

GURGLING CAVE - 14

Location: About 1000' east of Keyhole Rock

Entrances: 1

Length: 90.5'

Conditions: A cave which is best explored at low tide with calm seas. Not suitable for exploration by dinghy. It's dark inside, so a light is a necessity.

Description: A typical single-passage fissure cave formed on a north-south trending fault. The passage averages 5' wide and 9' high or so but gets up to 10' wide on the small sand beach in the rear, where the rise and fall of the swell created a soothing gurgling sound.

SELDOM SEEN CAVE - 15

Location: 1000' east of Keyhole Rock, and just to the left (east) of Gurgling Cave.

Entrances: 1

Length: 114'

Conditions: This cave can only be entered at low tide, as the entrance itself completely submerges. A light is also a necessity, as are calm seas. Exploration must be conducted on foot.

Description: The 10' high entrance soon drops to 4' high, then opens up into a 10' high fissure passage inside. This passage averages 2 to 3' wide and is water-floored for its entire length. Two-thirds of the way back the passage appears to end at a large rock, but one can sqeeze up and over this rock and continue another 43' to a sand beach.

Seldom Seen (on left) and Gurgling (on right) entrances

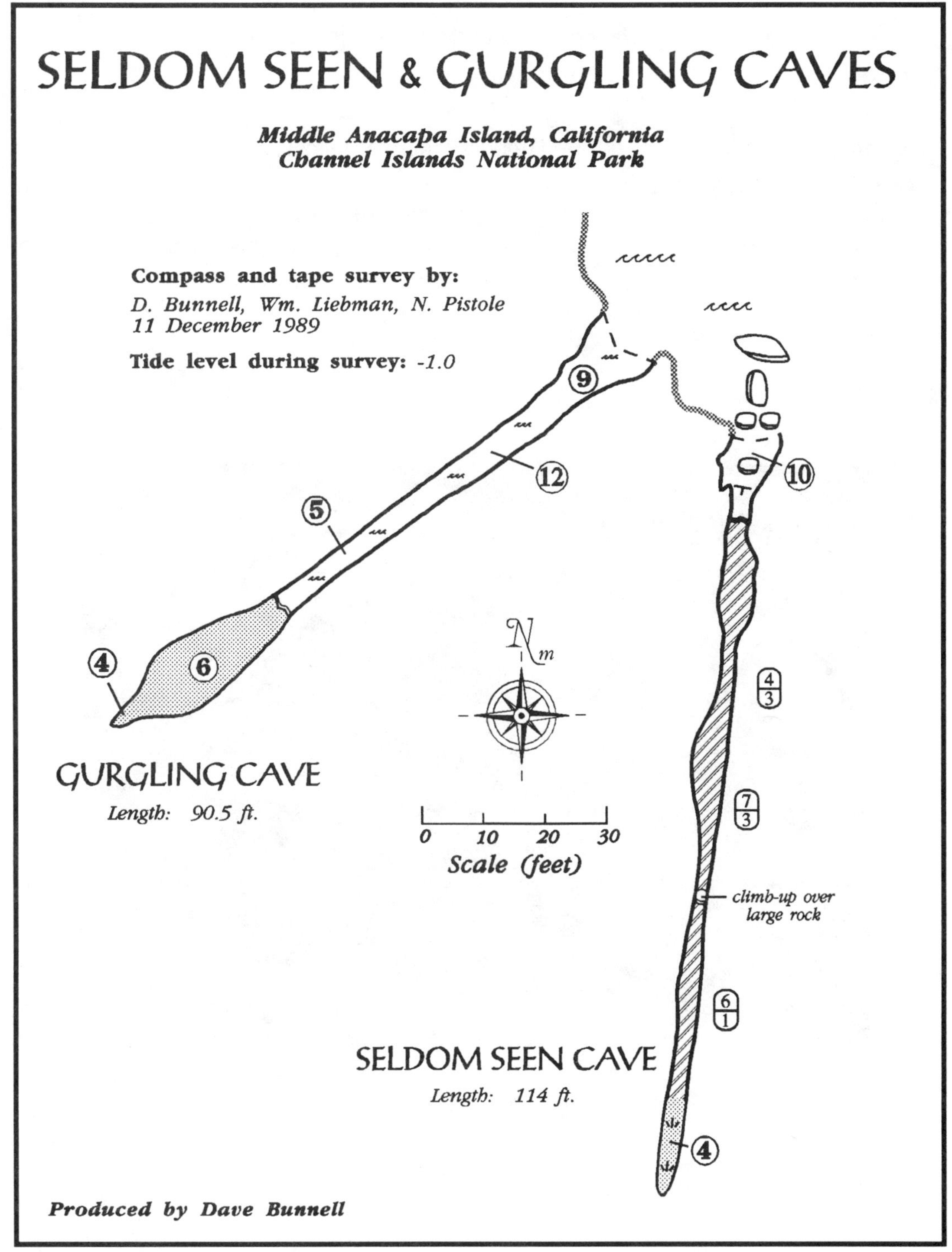
SELDOM SEEN & GURGLING CAVES
Middle Anacapa Island, California
Channel Islands National Park
Compass and tape survey by:
D. Bunnell, Wm. Liebman, N. Pistole
11 December 1989
Tide level during survey: -1.0
9
12
5
4
6
10
4/3
7/3
climb-up over large rock
6/1
4
GURGLING CAVE
Length: 90.5 ft.
0 10 20 30
Scale (feet)
SELDOM SEEN CAVE
Length: 114 ft.
Produced by Dave Bunnell

BATSTAR CAVE - 16

Location: Just west of the Park Service's Middle Island sign.

Entrances: 1

Length: 97'

Conditions: Explore at low tide and bring a light. Not suitable for exploration by dinghy.

Description: A standard single-passage fissure cave along a fault bearing 30-210°. Ceiling heights average about 10' and the passage varies from 4 to 10' wide. At low tide the floor is covered with some shallow tidepools and large cobbles. The cave ends in a sloping cobble beach.

Batstar Cave entrance at a moderately low tide

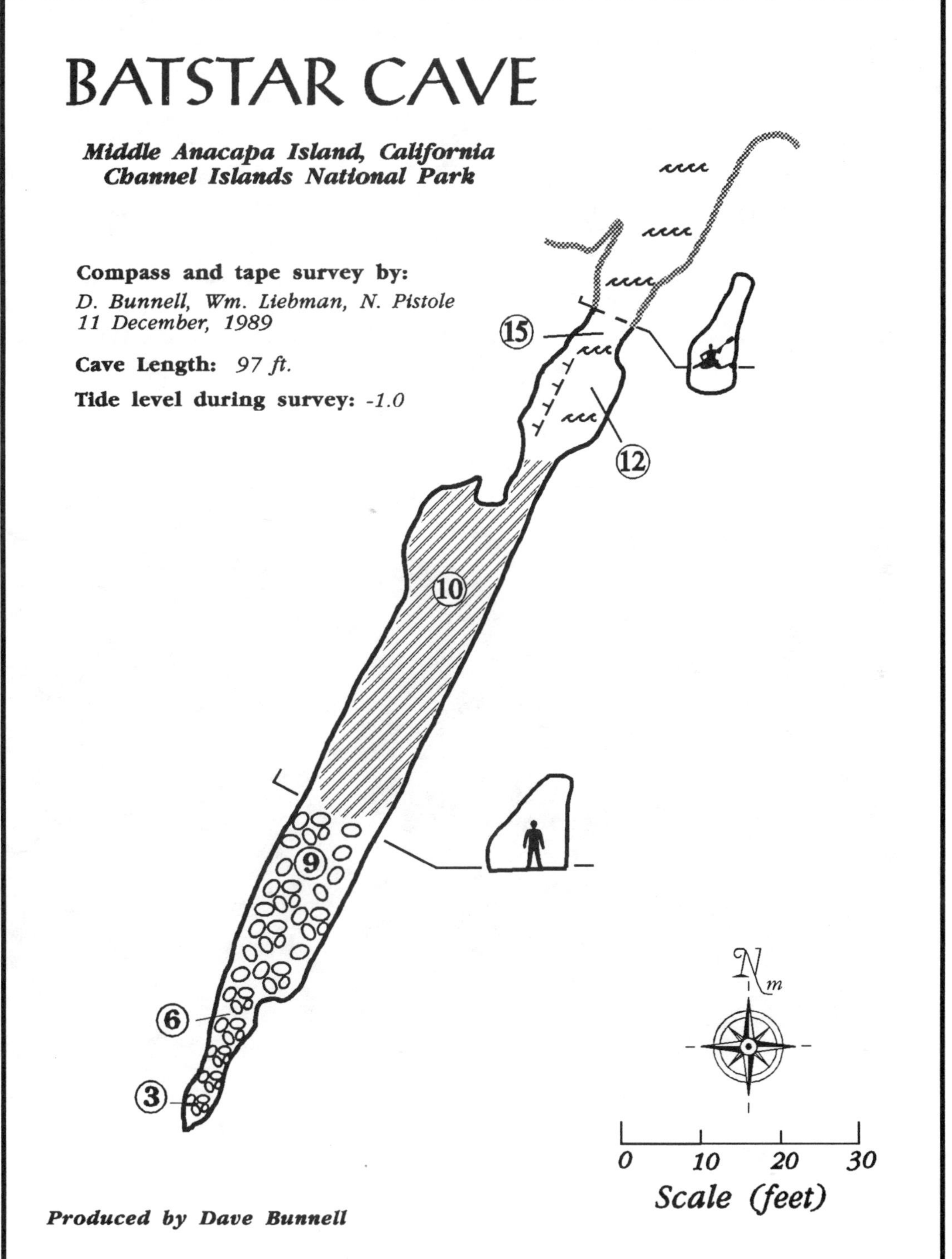
BATSTAR CAVE
Middle Anacapa Island, California
Channel Islands National Park
Compass and tape survey by:
D. Bunnell, Wm. Liebman, N. Pistole
11 December, 1989
Cave Length: 97 ft.
Tide level during survey: -1.0
15
12
10
9
6
3
N m
0 10 20 30
Scale (feet)
Produced by Dave Bunnell

AGATE BEACH CAVE - 17

Location: Beneath the park service's Middle Island sign. Shown on the USGS topographic map.

Entrances: 1

Length: 136'

Conditions: Best explored by swimming or wading in at low tide. However, ceiling heights are high enough that a moderate to high tide exploration is possible. A light is needed for the rear portions. The passage is too narrow to turn a kayak or dinghy.

Description: This cave was probably much longer at one time judging by the long surge channel leading into it. The cave's entrance is 30' high and 5' wide, somewhat keyhole shaped as viewed from the inside. The passage widens inside to about 10' wide and runs about 12' in height throughout. The cave ends at a beach where some agate cobbles were seen.

Agate Beach Cave on left; fissure on right doesn't go

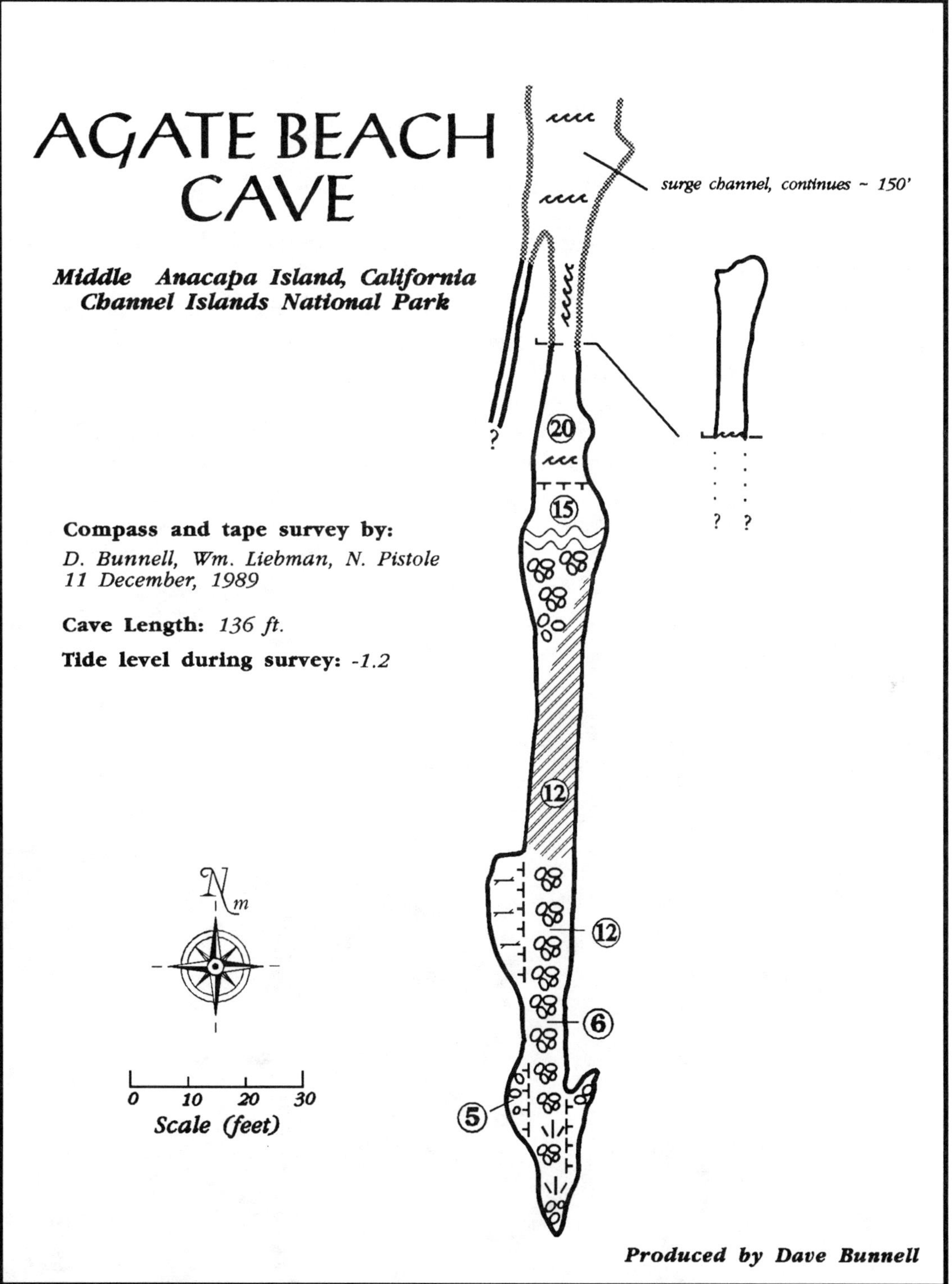
AGATE BEACH CAVE
Middle Anacapa Island, California
Channel Islands National Park
surge channel, continues ~ 150'
Compass and tape survey by:
D. Bunnell, Wm. Liebman, N. Pistole
11 December, 1989
Cave Length: 136 ft.
Tide level during survey: -1.2
20
15
12
12
6
5
? ?
?
N m
0 10 20 30
Scale (feet)
Produced by Dave Bunnell

COMPLEX CHASM - 18

Location: In a small cove, about 1500' east of Keyhole Rock. Just to the east of the entrance there is a large sea stack with two arches and a small tunnel.

Entrances: 2

Length: 390'

Conditions: Lights are needed to explore this large and impressive cave, which affords a possible through-trip in a kayak or small dinghy if conditions permit, ideally at a tide level of 1.0 or better to avoid rocks exposed at lower water. However, other sections of the cave open only at low tide.

Description: There are two entrances (see photos following the map). The easternmost is a 10' high classic fault pasage. The passage widens just inside, where there are numerous submerged urchin-covered rocks. About 45' inside the passage narrows to 3' but widens gradually beyond and opens into a wide room, where it intersects a passage coming from a second entrance. At low tide, most of this chamber was ankle deep and sand-floored. At the convergence, a left turn extends up to a broad sandy beach. Along the west wall of the main passage, a series of side passages are encountered, all following roughly parallel trends. The largest of these is just inside the second entrance and opens only at a very low tide. This side passage extends some 60' and ends in a small cobble beach.

Looking from the junction room towards the west entrance

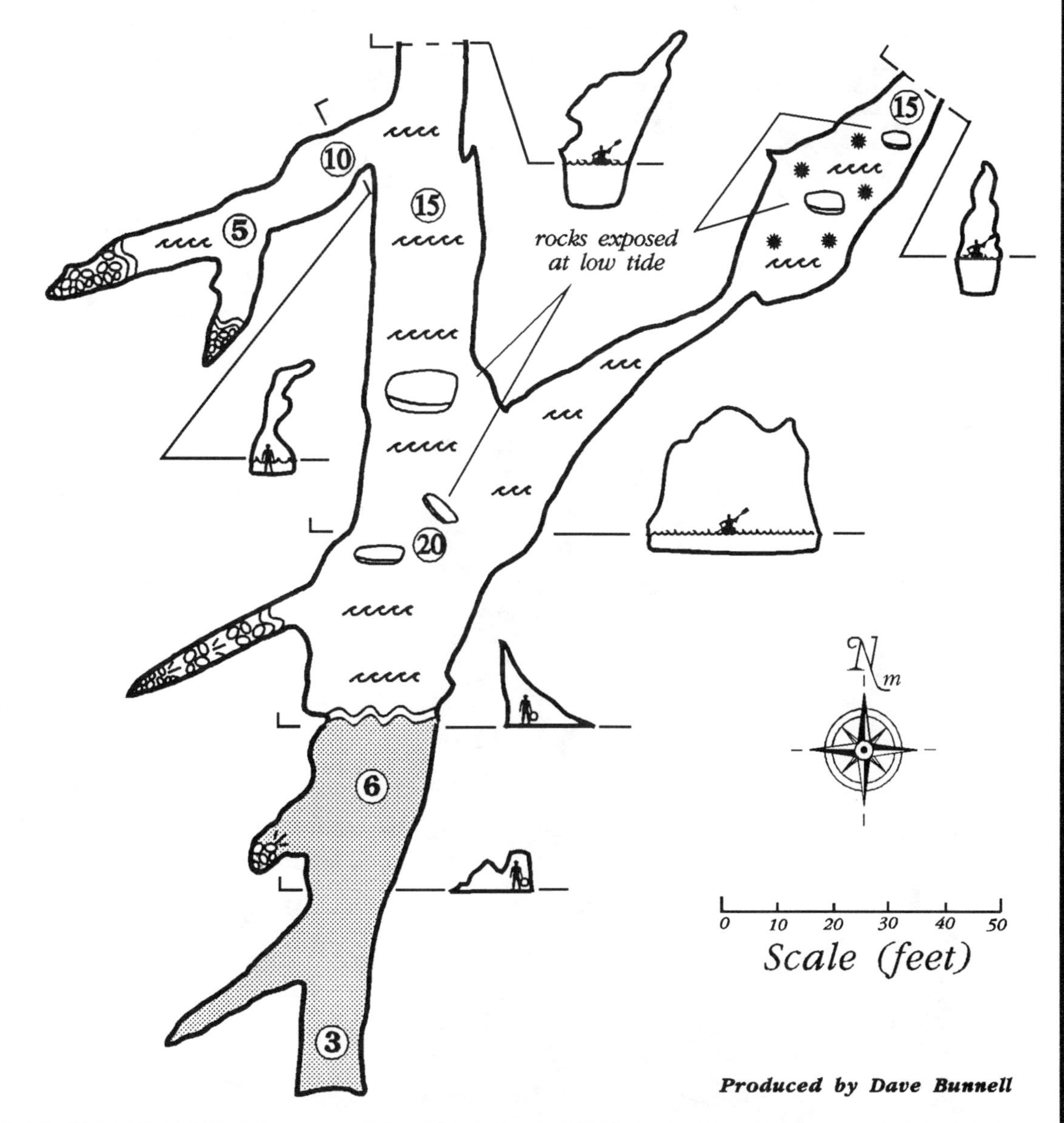
COMPLEX CHASM
Middle Anacapa Island, California
Channel Islands National Park
Compass and tape survey by:
D. Bewley, D. Bunnell
6 October 1990
Cave Length: 390 ft.
Tide level during survey: -0.1
15
10
15
5
rocks exposed
at low tide
20
6
3
N
m
0
10
20
30
40
50
Scale (feet)
Produced by Dave Bunnell

Eastern (left-hand) entrance to the Complex Chasm

Looking out the eastern entrance towards the large seastack that makes a good landmark for finding this cave

Western or right-hand entrance to the Complex Chasm

Looking through the arch that lies just to the left (east) of Converging Faults Cave, described on the next page

CONVERGING FAULTS CAVE - 19

Location: ~ .5 mile west of the east tip of Middle Island, north shore. It is indicated on the USGS topo map and is partially concealed by a large sea stack.

Entrances: 1

Length: 133'

Conditions: Explorable in a dinghy or kayak if sea conditions are calm. A light is useful in the rear portions.

Description: The entrance is some 40' wide and 8' high. The two walls are each formed along separate faults which converge at the rear of the cave. The entrance chamber extends some 80' to a sand beach which slopes and narrows to a squeeze through cobbles. Beyond there is another 20' of passage.

Converging Faults Cave entrance

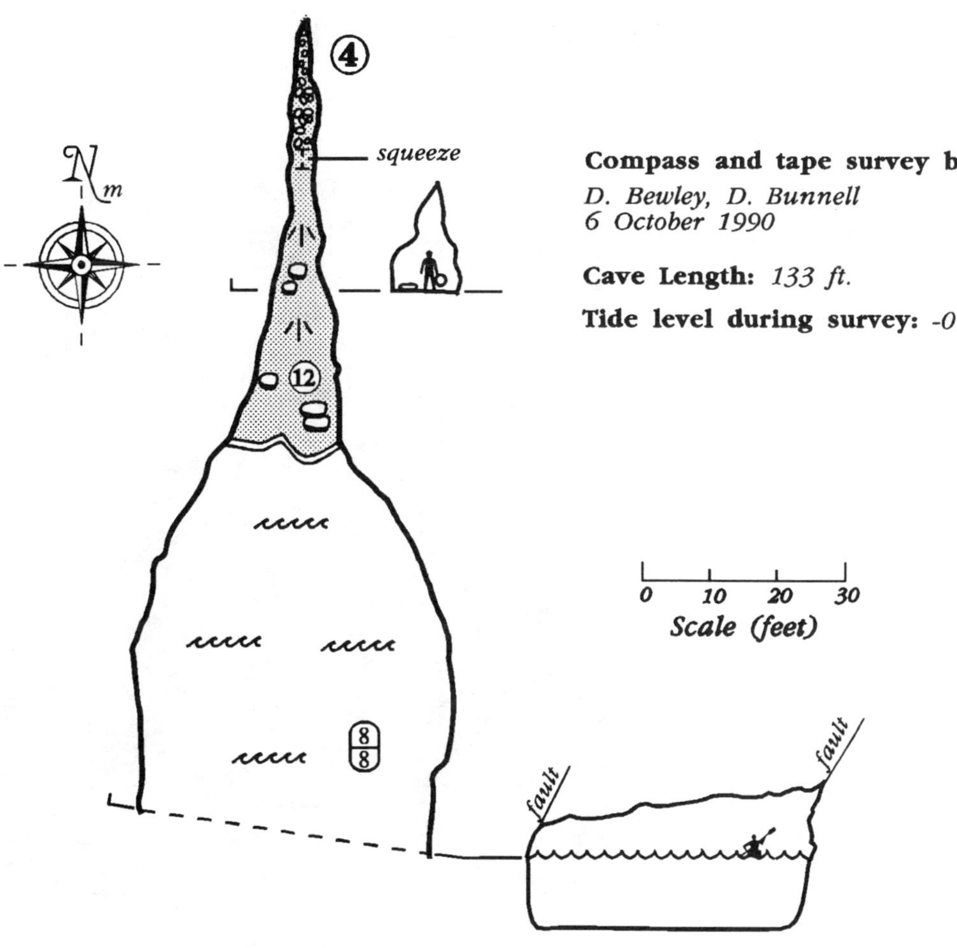
CONVERGING FAULTS CAVE
Middle Anacapa Island, California
Channel Islands National Park
N m
squeeze
4
12
8
8
Compass and tape survey by:
D. Bewley, D. Bunnell
6 October 1990
Cave Length: 133 ft.
Tide level during survey: -0.1
0 10 20 30
Scale (feet)
fault
fault
Produced by Dave Bunnell

LITTLE SURPRISE CAVE - 20

Location: 2000' west of the east tip of Middle Island, on the west side of a small promontory.

Entrances: 2

Length: 130'

Conditions: Completely dry at low tide; no lights are needed.

Description: A small cave with a sloping cobble floor which surprised us by the presence of a side passage leading to a second entrance.

EL CHIQUITO - 21

Location: Just north of Little Surprise Cave

Entrances: 1

Length: 34.7'

Conditions: Dry at low tide; no lights are needed.

Description: An arched entrance 8' wide and 10' high leads into a short passage.

Ernie Garza enters the small side entrance of Little Surprise Cave

LITTLE SURPRISE & EL CHIQUITO CAVES

Middle Anacapa Island, California
Channel Islands National Park

Compass and tape survey by:
D. Bunnell, E. Garza, M. Oliphant
15 October 1989

Tide level during survey: *-0.9*

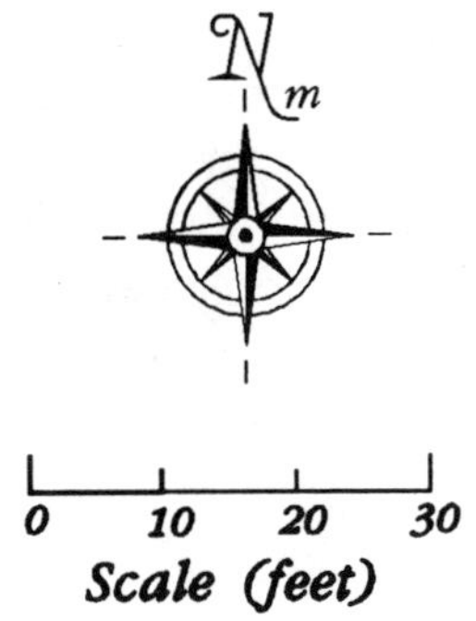

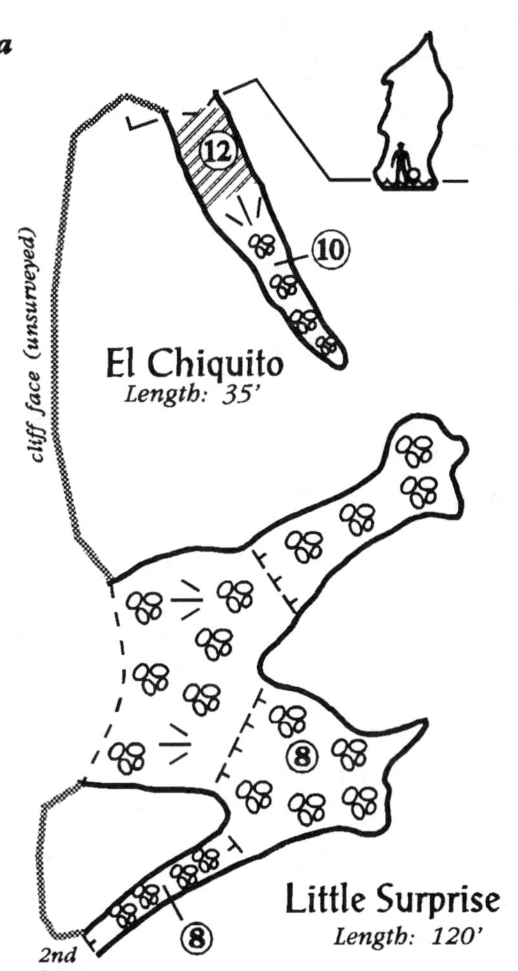

Produced by Dave Bunnell

TREASURE CHEST CAVE GROUP

Location: This group of caves are centered on a small promontory, about 2000' west of the east tip of Middle Island.

WESTVIEW CAVE - 22

Entrances: 1

Length: 82'

Conditions: Mostly dry at low tide and explorable at any tide level. No lights are needed. Rocks in front render the landing of a dinghy problematic in any but the calmest of seas.

Description: A large 30' high x 40' wide entrance leads into a hemisperically-shaped room with a sloping cobble floor.

TREASURE CHEST CAVE - 23

Entrances: 4

Length: 207'

Conditions: At a very low tide, much of the cave empties out and one must explore on foot. At high tide, the cave makes a nice through-trip in a kayak or dinghy, sea conditions permitting. Lights are needed only in the "treasure chest" section of the cave.

Description: The short length of this cave belies its interest and complexity. The cave has four passages which radiate from two drip lines, lying at right angles. The complex is named for a rectanglular metal box seen embedded in the rock in the right-hand tunnel. The box and much of the passage is floored with multicolored sponge and algal growth: a beautiful spot at low tide. This passage is largely walking height and about 60' long. Next to this is an impressive passage which cuts 75' through the promontory. It is 15-20' wide and 8-15' high and floored with broken rock. To the west is a dripline which parallels that which formed this tunnel. Just past this dripline is a passage 28' long averaging 7' high. The main passage on this side cuts 50' through the northwest corner of the promontory and is water-floored throughout.

VICTORY CAVE - 24

Entrances: 2

Length: 183'

Conditions: Largely water-floored, even at low tide. The western entrance could be entered by dinghy in calm seas. No lights are needed.

Description: Two tunnels lead into a moderate-sized chamber with a sloping cobble beach. The western entrance is water-floored while the eastern entrance is dry at low tide, with fallen rocks in the dripline area.

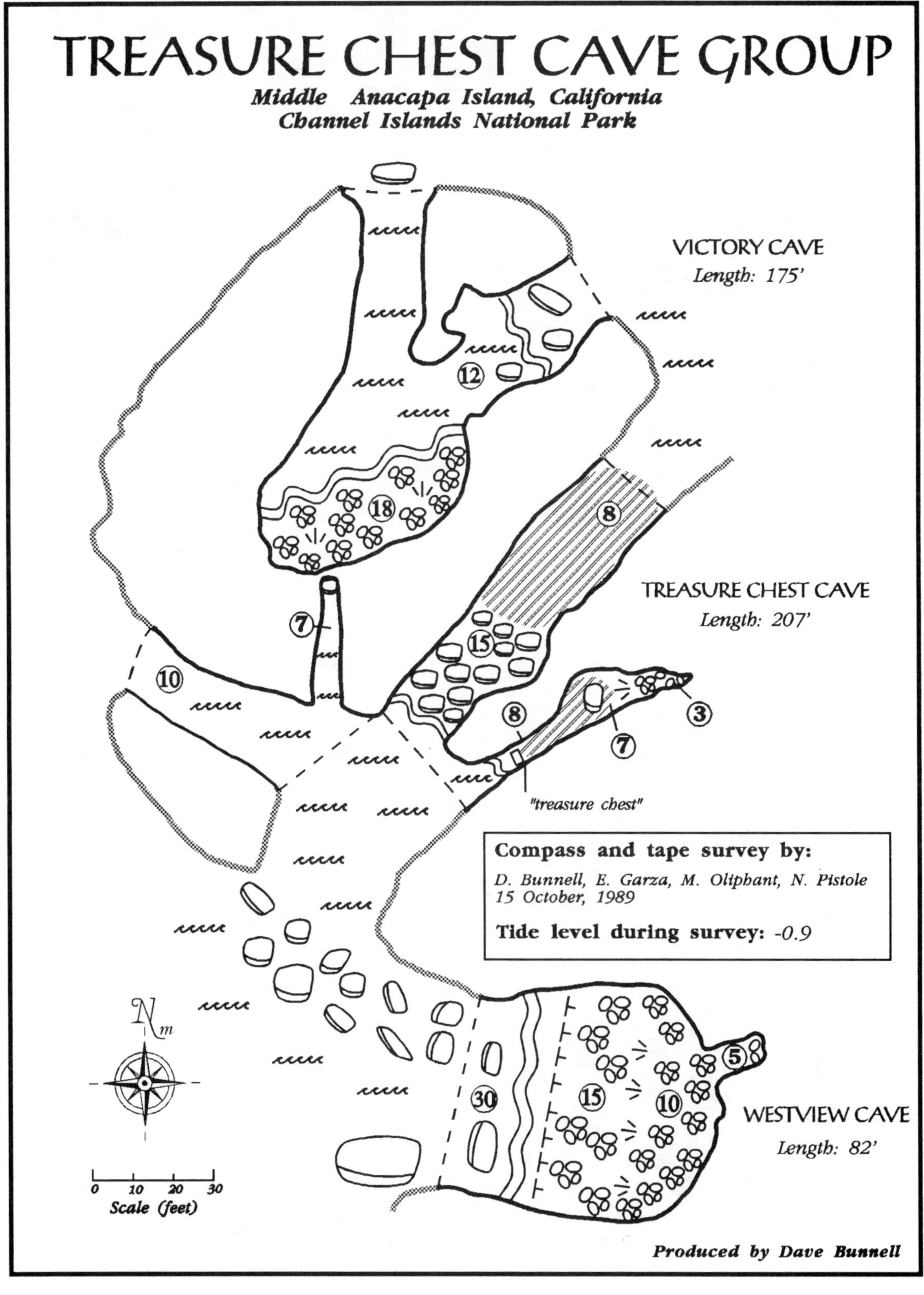
TREASURE CHEST CAVE GROUP
Middle Anacapa Island, California
Channel Islands National Park
VICTORY CAVE
Length: 175'
TREASURE CHEST CAVE
Length: 207'
"treasure chest"
Compass and tape survey by:
D. Bunnell, E. Garza, M. Oliphant, N. Pistole
15 October, 1989
Tide level during survey: -0.9
WESTVIEW CAVE
Length: 82'
0 10 20 30
Scale (feet)
Produced by Dave Bunnell

Looking out of Westview Cave towards West Anacapa

Treasure Chest Cave entrances on SW side of point: tunnel on left, single entrance "treasure chest" section on right

Treasure Chest Cave entrances on the west side of the point

Treasure Chest Cave, entrance on NE side of point. Note the two parallel faults, giving the passage a rectangular cross-section instead of the typical vaulted ceiling formed along single faults.

CRAWLSPOT CAVE - 26

Location: About 600' from the east end of Middle Island

Entrances: 1

Length: 195'

Conditions: The cave is generally dry and mostly cobble-floored. A light is necessary beyond the constriction.

Description: The cave at first appears to be a simple chamber like its neighbor, Sloping Cobble. However, a passage on the right leads back 60' beyond an "A" -shaped constriction, opening to a walking-height fissure floored with angular, broken rock.

The entrance to Crawlspot Cave, just next to Sloping Cobble

CRAWLSPOT CAVE

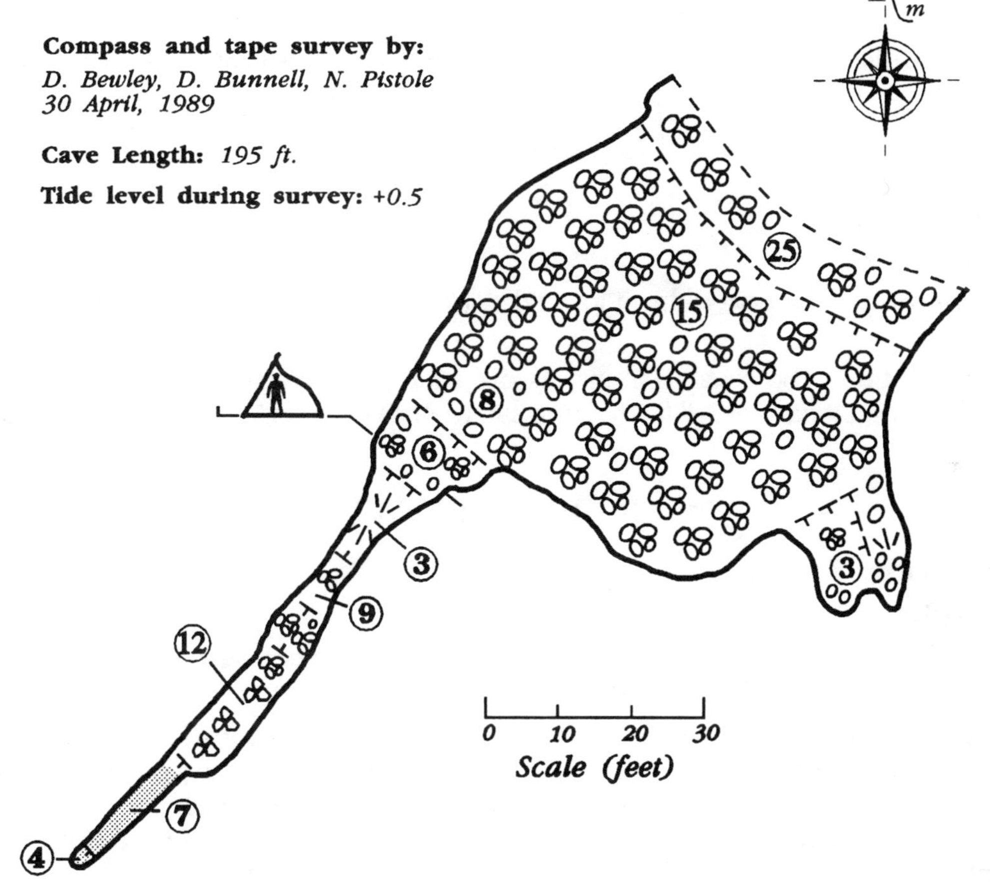

Produced by Dave Bunnell

EAST END CAVE GROUP

Location: This group of 3 caves is about 400' from the east end of Middle Island in a small cove.

SLOPING COBBLE CAVE - 27

Entrances: 1

Length: 56'

Conditions: The cave is generally dry and mostly cobble-floored. No light is needed and the whole cave is visible without landing.

Description: The cave is a single cobble-floored chamber with an entrance 45' wide and 15' high. The floor slopes up to the back wall of the cave.

SAND CAVE - 27

Entrances: 1

Length: 94'

Conditions: The cave is dry at low tides. No lights are needed except in the very rear portions.

Description: The cave's entrance is 40' wide and 15' high. The cave is a single, walking-height passage, all floored with sand, which narrows to 4' wide and high towards the rear.

UNDERCUT CAVE - 28

Entrances: 2

Length: 119'

Conditions: The cave should be explored at low tide. Some crawling is required to do the "through-trip", which can be done without a light.

Description: The cave has two entrances which connect through a low, undercut portion of the wall which once separated them. The left-hand entrance is partially waterfloored at low tide and leads into a vaulted, walking height passage which penetrates 85' into the cliff. Some rusting lobster traps were seen in the rear. The right-hand entrance leads into a sand-floored room which connects into the main passage to the left.

From left to right, these are the entrances to Undercut, Sand, and Sloping Cobble Caves

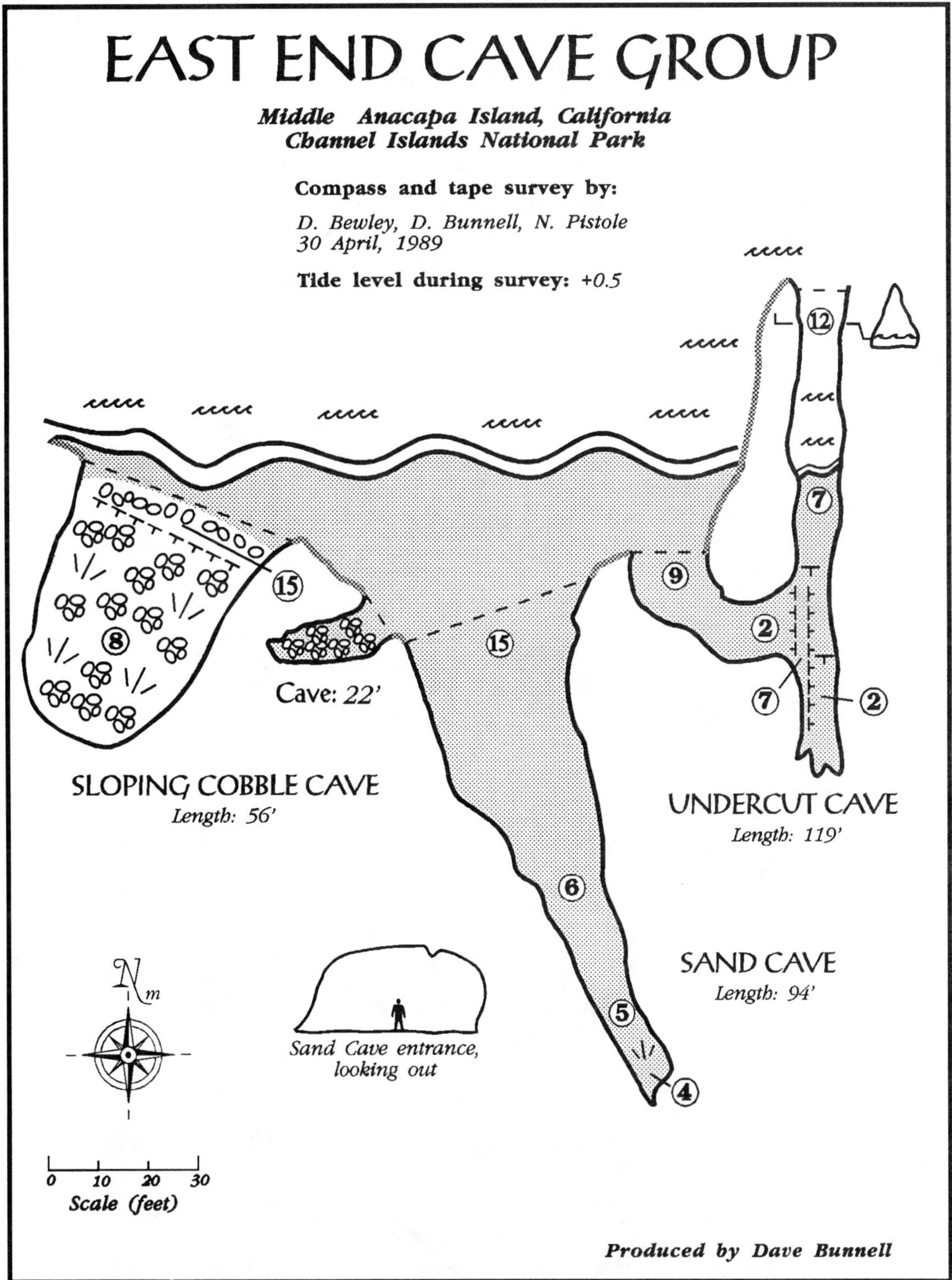
EAST END CAVE GROUP
Middle Anacapa Island, California
Channel Islands National Park
Compass and tape survey by:
D. Bewley, D. Bunnell, N. Pistole
30 April, 1989
Tide level during survey: +0.5
Cave: 22'
SLOPING COBBLE CAVE
Length: 56'
UNDERCUT CAVE
Length: 119'
SAND CAVE
Length: 94'
Sand Cave entrance, looking out
0 10 20 30
Scale (feet)
Produced by Dave Bunnell

SEE-THE-STACK CAVE - 29

Location: In a small, protected cove on the SE tip of Middle Island. A cobble beach is just to the right of the entrance.

Entrances: 1

Length: 73'

Conditions: The cave empties out completely at low tide, so this cave must be explored on foot. No lights are needed. Can't be entered in a dinghy.

Description: The cave's entrance is 25' high and 18' wide. A large breakdown block is seen at the dripline. The cave is a single chamber with sloping cobble floor leading into a short passage . Several large sea stacks are visible from the entrance, affording a nice photo opportunity.

See-the-Stack Cave is formed along a prominent fault

SEE-THE-STACK CAVE

Middle Anacapa Island, California
Channel Islands National Park

Compass and tape survey by:
D. Bewley, D. Bunnell, D. Morris
8 June 1992

Cave Length: *73 ft.*

Tide level during survey: *0.8*

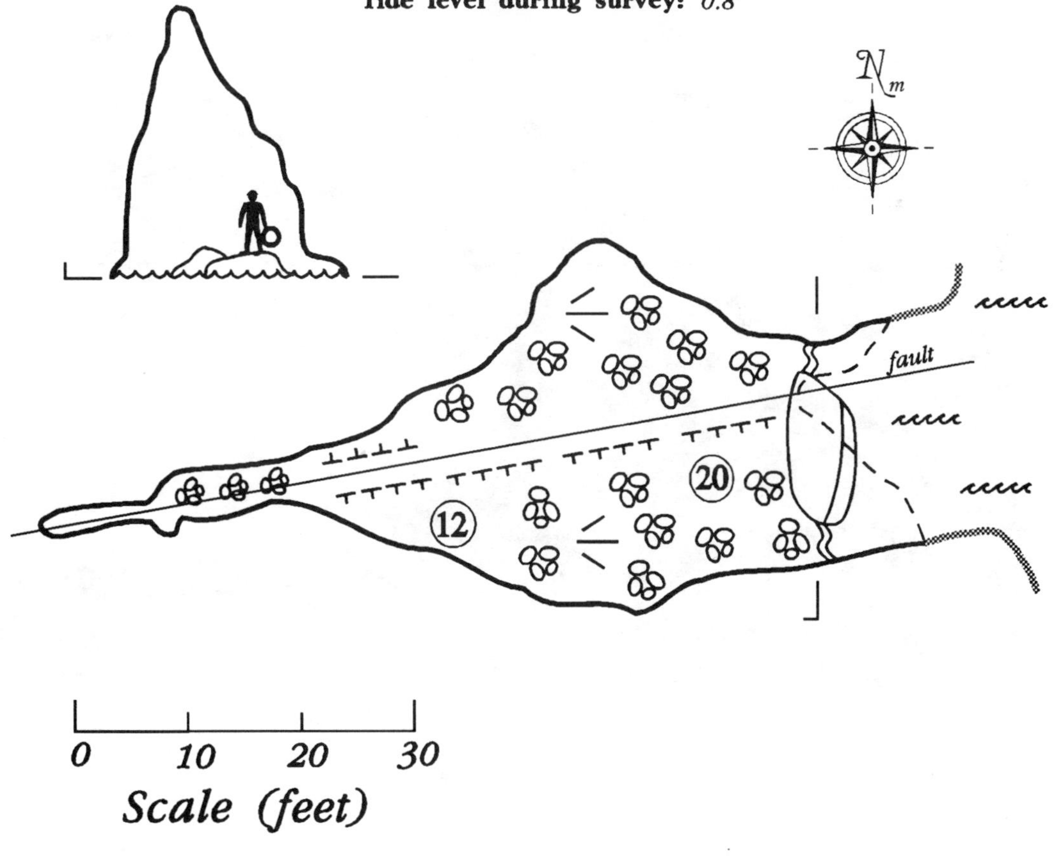

Produced by Dave Bunnell

MORNING THUNDER CAVE - 30

Location: approximately 1000' west of the east tip of Middle Island, on the south shore

Entrances: 1

Length: 130'

Conditions: Can be explored by a dinghy for its entire length if seas are calm. A light would be useful in the rear portions.

Description: One of the few "deep-water" caves on the south shore. The 20'x 20' entrance leads through water-floored passage. There is no beach to land on except at very low tides. Ceiling heights are seven to eight feet and higher.

Kayaking by the Morning Thunder entrance

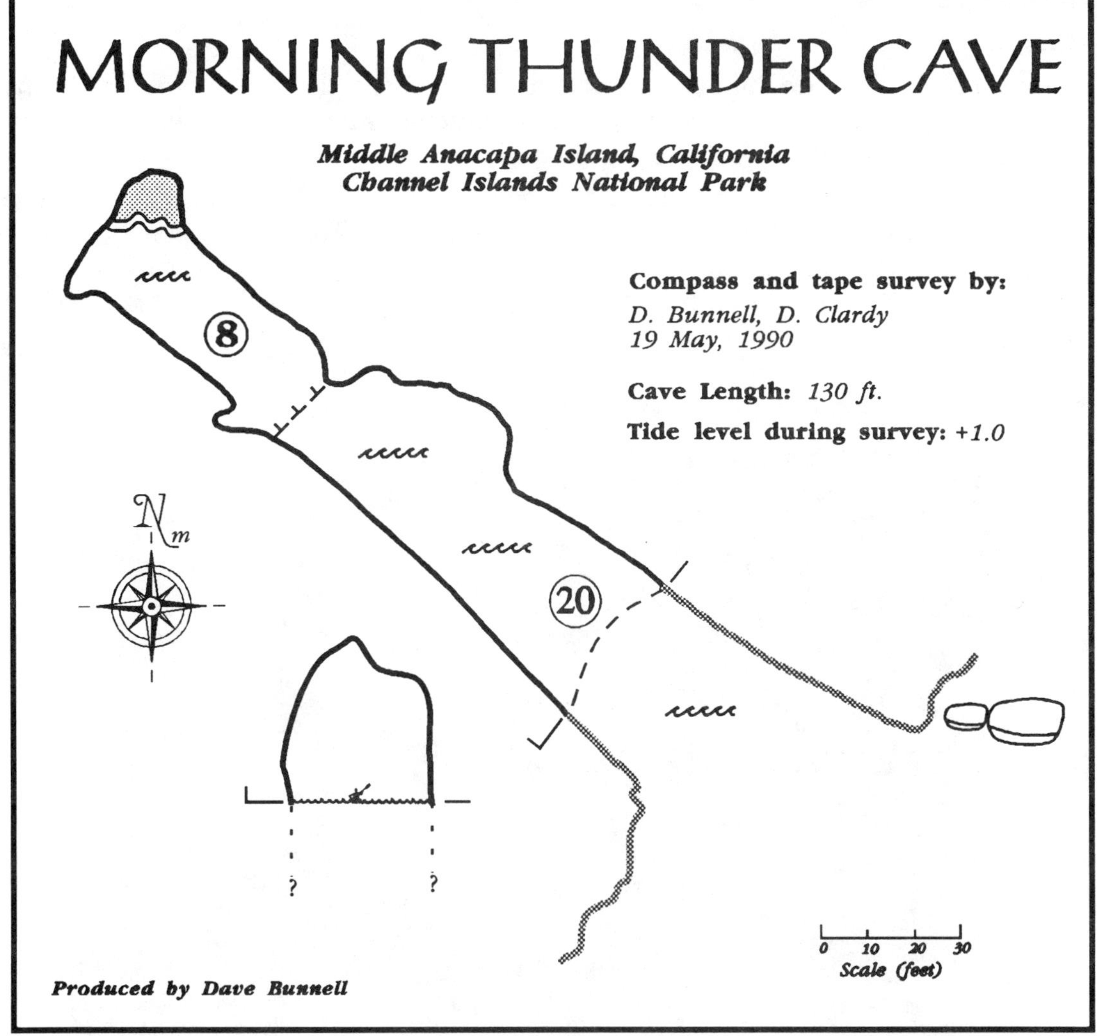
MORNING THUNDER CAVE
Middle Anacapa Island, California
Channel Islands National Park
Compass and tape survey by:
D. Bunnell, D. Clardy
19 May, 1990
Cave Length: 130 ft.
Tide level during survey: +1.0
8
20
?
?
0 10 20 30
Scale (feet)
Produced by Dave Bunnell

REFUGE CAVE - 31

Location: approximately 2000' west of the east tip of Middle Island, on the south shore. Unmapped cave features, consisting of alcoves on the same level, are seen on both sides of the entrance (2 to the right, 1 to the left).

Entrances: 1

Length: 54'

Conditions: The cave must be explored on foot; it is completely dry (probably) at any tide. At low tide one might land on the cobble beach in front of the cave. We tied our kayaks up to the shelf above. No lights are needed.

Description: The cave's entrance is 40' wide and 20' high. One climbs up some five feet into the cave, which is floored with the lava bedrock. There are a couple of large breakdown blocks and the walls on the left side are composed of layered cobbles. Interesting debris from various shipwrecks was found at the base of the shelf leading into the cave.

Refuge Cave is on the left; on the right is a small (less than 20' long) unsurveyed cave

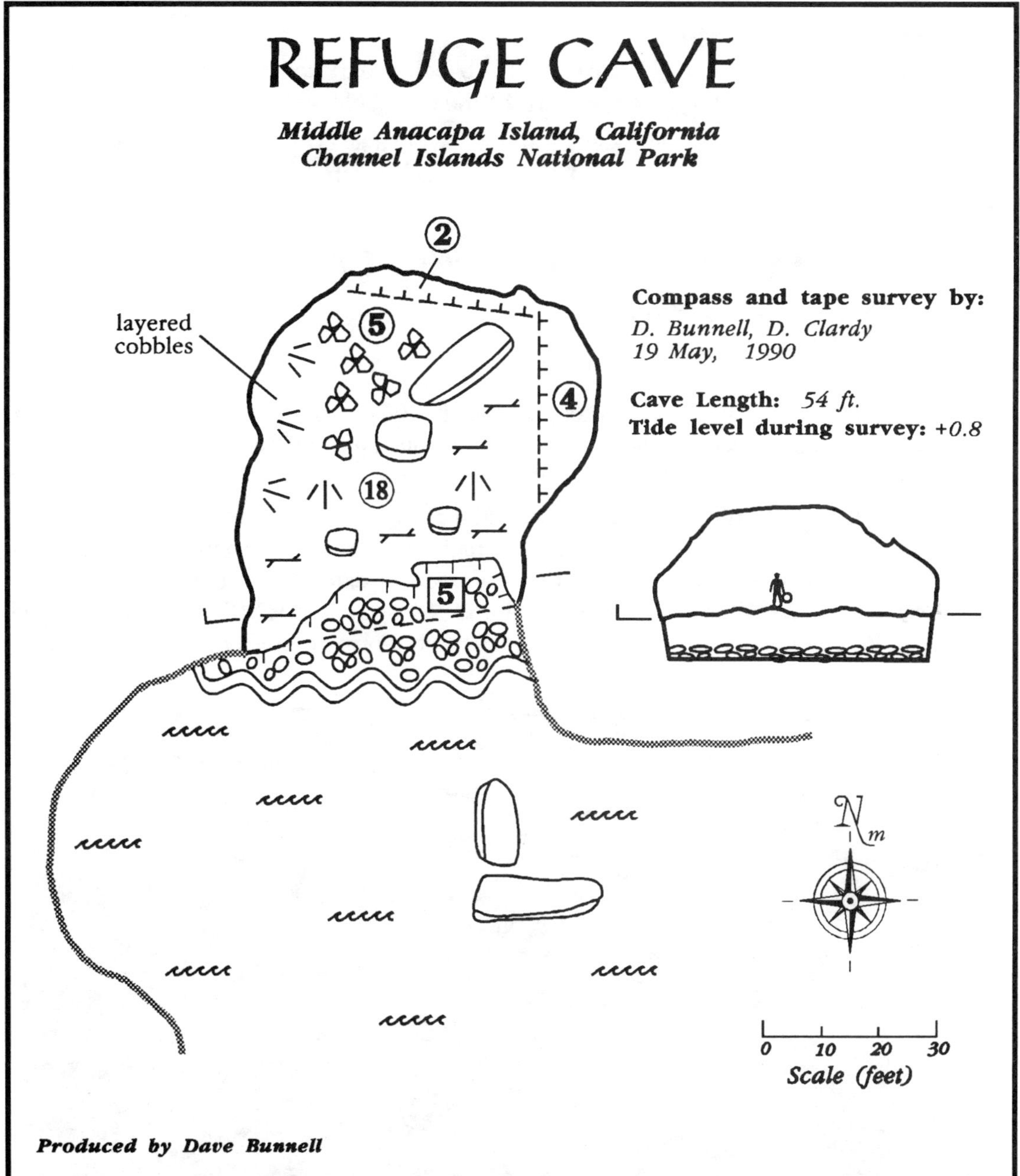
REFUGE CAVE
Middle Anacapa Island, California
Channel Islands National Park
Compass and tape survey by:
D. Bunnell, D. Clardy
19 May, 1990
Cave Length: 54 ft.
Tide level during survey: +0.8
layered cobbles
2
4
5
18
N
m
0 10 20 30
Scale (feet)
Produced by Dave Bunnell

LADIES' DELIGHT CAVE - 32

Location: Eastern edge of East Fish Camp

Length: 72'

Entrances: 1

Conditions: Water-floored. A small cobble beach is exposed at low tide. A kayak or dinghy might be used in the outer portions of the cave before it narrows.

Description: The cave's entrance is some 20' wide and 20' high. The passage curves to the left, narrowing halfway back to about 10' wide, with 12' high ceilings. The cave ends in a cobble beach.

HONEYCOMB WORM CAVE - 33

Length: 114'

Entrances: 1

Conditions: Water-floored. Two cobble beachs are exposed at low tide. A kayak or dinghy might be used in the outer portions of the cave before it narrows, although there are numerous submerged rocks. A light isn't necessary

Description: The entrance is 11' wide and 18' high. For the first 35', the cave is a single passage which gets up to 15' wide and high. It then splits into two branches. The left-hand passage is 22' long, and averages 4' wide. It extends through nicely sculpted walkway to a cobble beach. The wider right-hand passage extends 50' to a cobble-floored chamber 12' high and wide. Numerous honeycomb worm colonies coat the left wall of the cave.

Ladies' Delight on the right, Honeycomb Worm on the left

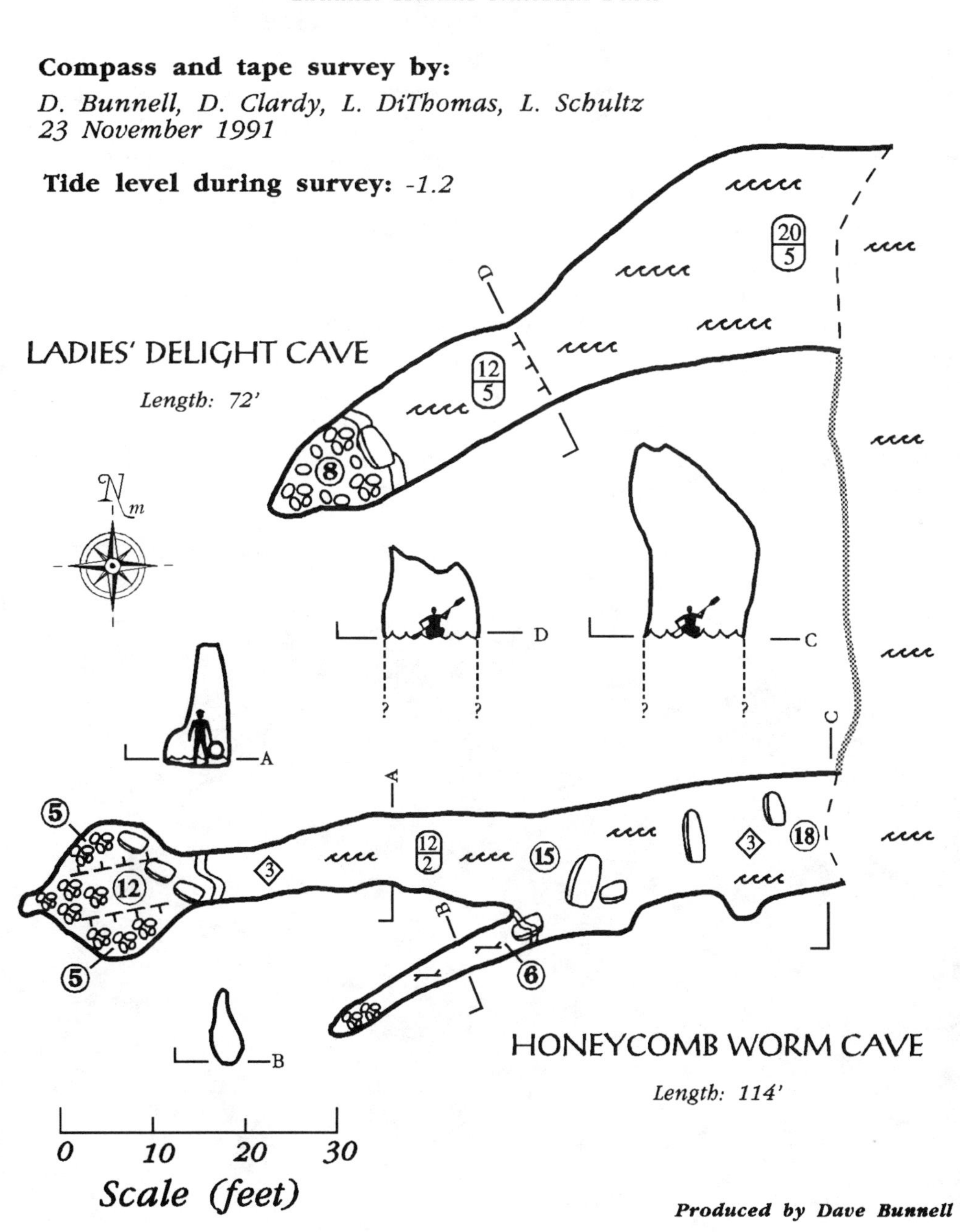
HONEYCOMB WORM & LADIES' DELIGHT CAVES
Middle Anacapa Island, California
Channel Islands National Park
Compass and tape survey by:
D. Bunnell, D. Clardy, L. DiThomas, L. Schultz
23 November 1991
Tide level during survey: -1.2
LADIES' DELIGHT CAVE
Length: 72'
HONEYCOMB WORM CAVE
Length: 114'
0 10 20 30
Scale (feet)
Produced by Dave Bunnell

SHIPWRECK CAVE - 34

Location: East Fish Camp region

Length: 81'

Entrances: 1

Conditions: The cave is largely dry at low tide and entry by swimming or dinghy is probably easiest. We originally entered by climbing down the cliff-face on the left side of the entrance (see photo). However, there are numerous rocks and metallic debris, so calm seas are advised before attempting a landing here.

Description: This is a short but impressive cave. Much of the outer portions of the cave are bedrock-floored, partially covered by numerous fallen rocks. On the west wall there are nicely sculpted alcoves, encrusted with honeycomb worm colonies. The cave narrows to 25' some 40' from the dripline, where the floor is sand and cobble-covered. Debris from a shipwreck lines the cave, most prominently a long propeller shaft, some 25' long, and a large propeller, estimated to weigh 750 pounds (see photos on the next two pages). According to Park Service archaeologoist Don Morris, this is from the wreck of the wooden fishing vessel *Equator*, which went down in 1949. The rest of the vessel lies in about 20' of water just offshore from the cave. During a return visit on October 25 (11 months after our survey) we found that additional material had moved in to the cave. Currently, the Park Service is deciding whether to remove the prop for display somewhere, and if so, how. In any event, this wreckage, like any other in the Park, is protected by law. Leave the cave and its contents as you found it!

Kayaking towards Shipwreck Cave; beware of rocks!

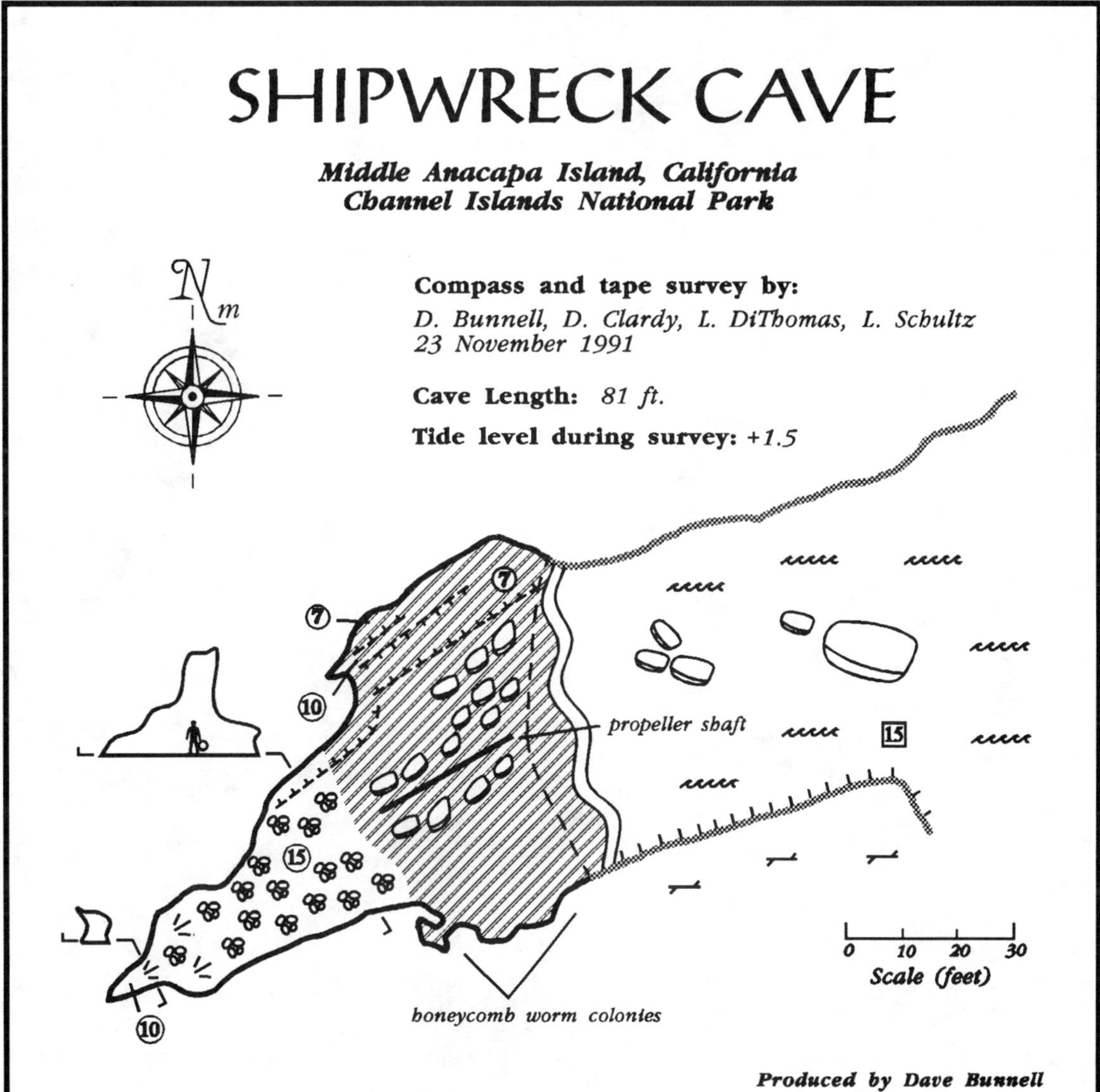
SHIPWRECK CAVE
Middle Anacapa Island, California
Channel Islands National Park
N
m
Compass and tape survey by:
D. Bunnell, D. Clardy, L. DiThomas, L. Schultz
23 November 1991
Cave Length: 81 ft.
Tide level during survey: +1.5
7
7
10
15
15
10
propeller shaft
0 10 20 30
Scale (feet)
honeycomb worm colonies
Produced by Dave Bunnell

Looking out from Shipwreck Cave. The shaft is over 25' long.

Park Service archaeologist Don Morris examines the prop washed up in Shipwreck Cave

Carl Reuter examines the 25' long propeller shaft

GREEN ABALONE CAVES - 35

Location: Eastern edge of East Fish Camp anchorage

Length: 33 & 27'

Entrances: 1

Conditions: Best explored at low tide by swimming in. The caves are dry then.

Description: These caves barely meet our survey standards, but are included because they have such prominent entrances and are closely grouped. There are two caves in the rear of this tiny cove in East Fish Camp. The lefthand cave is a cobble-floored chamber, 12' high and 27' deep. To its right is a deeper cave, floored with bedrock in its outer portion and with unusually large cobbles in its rear. The caves are formed in a small cliff topped by non-volcanic sediments and don't seem to be associated with any obvious faults.

The two Green Abalone Caves empty out at low tide

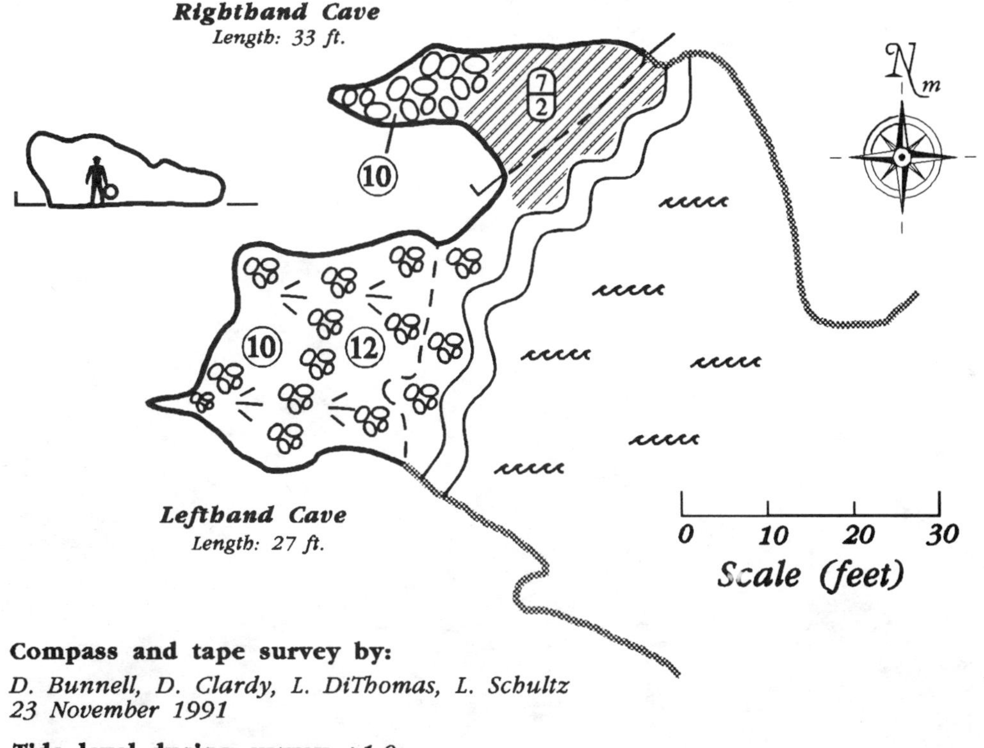
GREEN ABALONE CAVES
Middle Anacapa Island, California
Channel Islands National Park
Rightband Cave
Length: 33 ft.
Lefthand Cave
Length: 27 ft.
7/2
10
10
12
N m
0 10 20 30
Scale (feet)
Compass and tape survey by:
D. Bunnell, D. Clardy, L. DiThomas, L. Schultz
23 November 1991
Tide level during survey: +1.0
Produced by Dave Bunnell

FISH CAMP CAVES - 36 & 37

Location: East Fish Camp

Length: 74' and 55'

Entrances: each has only 1

Conditions: These caves empty out at low tide and can be explored on foot. Their floors are probably only water-covered at higher tides. The area around the caves is well protected from west swell. If landing a dinghy or kayak in this area, beware of urchins and rocks. No lights are needed.

Description: Numerous small caves are clustered together in this area, only two of which meet our 30' length criterion for a cave. Both caves have sandy floors covered with cobbles in some areas. The longer cave has the tallest entrance, some 25' high. A swallow's nest was seen on the ceiling of one cave.

The Fish Camp Caves. The largest one is the entrance on the far right. Note person on shore for scale.

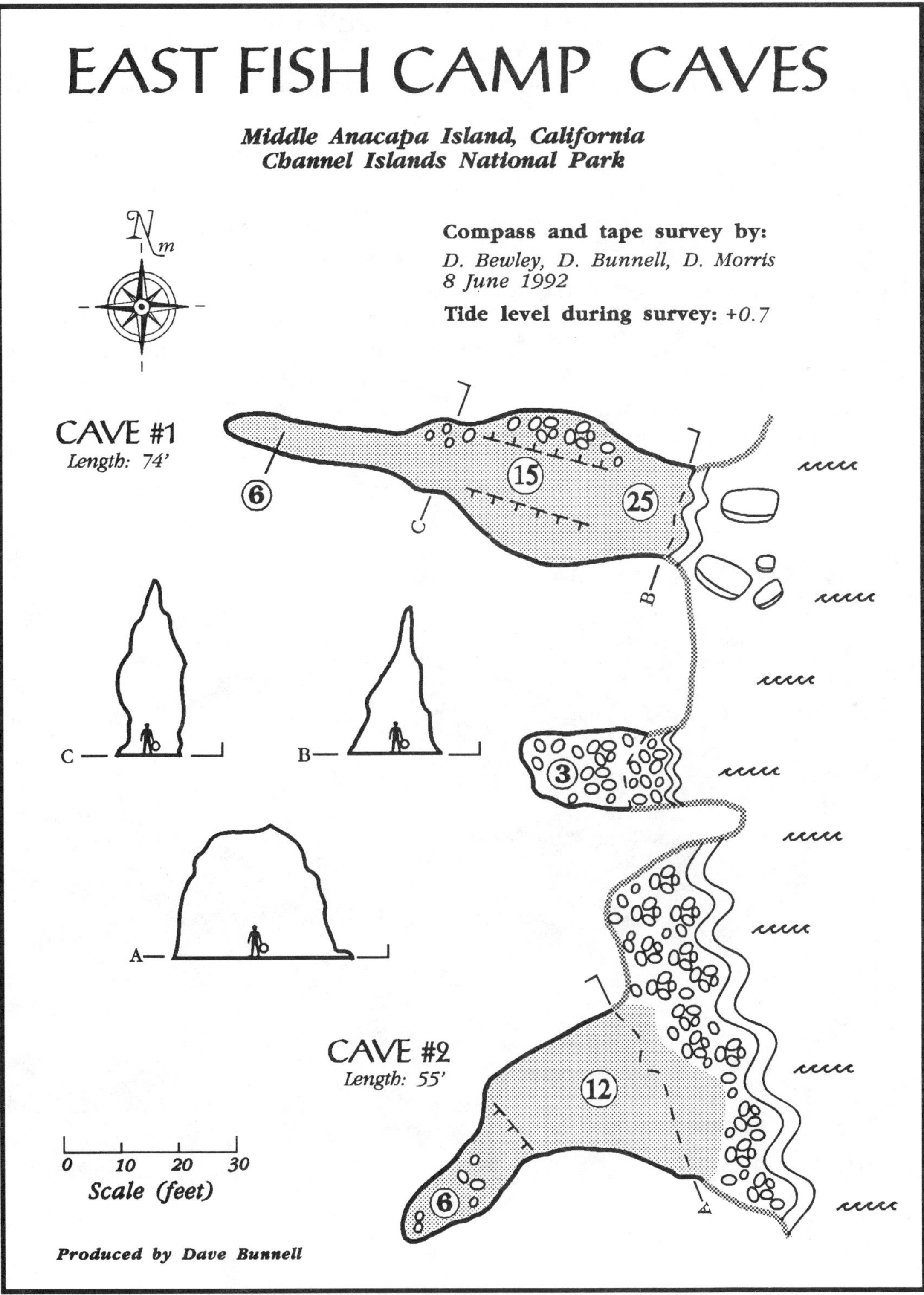
EAST FISH CAMP CAVES
Middle Anacapa Island, California
Channel Islands National Park
Compass and tape survey by:
D. Bewley, D. Bunnell, D. Morris
8 June 1992
Tide level during survey: +0.7
CAVE #1
Length: 74'
CAVE #2
Length: 55'
0 10 20 30
Scale (feet)
Produced by Dave Bunnell

LAVA BENCH CAVE #1 - 38

Location: In the cliffs above a large lava bench approximately 3/4 mile west of the East Fish Camp anchorage

Entrances: 1

Length: 58'

Conditions: This cave may be visited in virtually any tide level so long as conditions permit a landing, either by dinghy or swimming. Daylight illuminates the entire cave.

Description: This is a large, dry, upper level shelter cave which sits some 20' above the sea. The entrance is 40' wide and 25' high and leads into a chamber 58' deep and 40' wide. The front portions are flat bedrock, grading to broken rock and then a steep slope of sand and rock in the rear. The ceiling is quite irregular and in places the chunky lava is pendant-like. There is a trench below the main fault line which shows evidence of an intrusive dike - a feature not often seen in the caves here.

Lava Bench Cave #1 is high and dry.

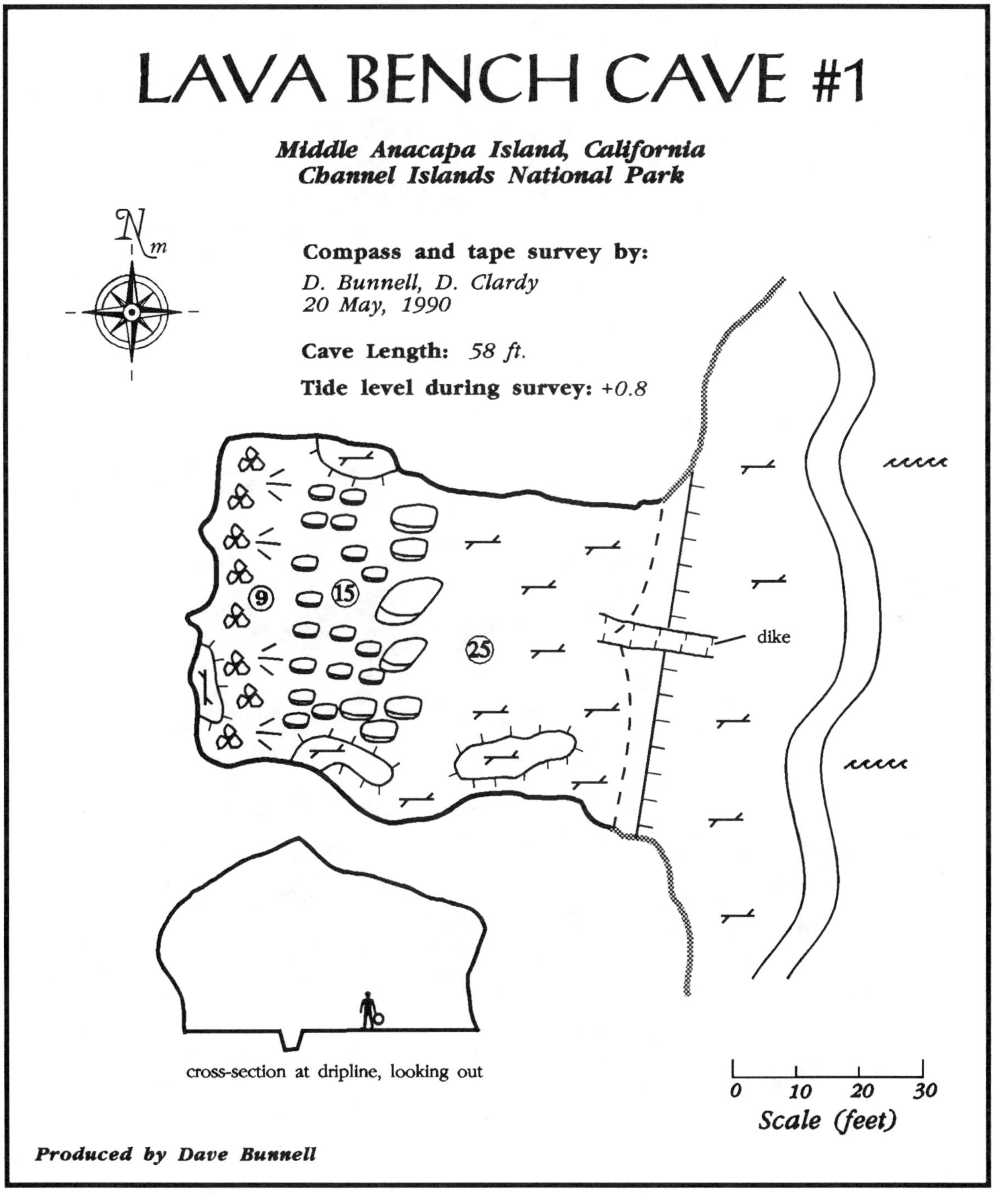
LAVA BENCH CAVE #1
Middle Anacapa Island, California
Channel Islands National Park
N m
Compass and tape survey by:
D. Bunnell, D. Clardy
20 May, 1990
Cave Length: 58 ft.
Tide level during survey: +0.8
9
15
25
dike
cross-section at dripline, looking out
0 10 20 30
Scale (feet)
Produced by Dave Bunnell

LAVA BENCH CAVE #2 - 39

Location: In the cliffs above a large lava bench approximately 3/4 mile west of the East Fish Camp anchorage

Entrances: 2

Length: 115'

Conditions: The cave is completely dry at any tide level. One can land a dinghy on the lava bench (there is plenty of room to beach it) if the surge permits. The climb into the cave is easy but a bit exposed. No light is necessary to explore the cave.

Description: This is another two-level cave, but the lower level is unenterable (except perhaps with scuba). The entrance is about 40' wide and 50' high down to water level. The floor is all bedrock, with unusual fin-like projections. Two-thirds of the way back, the floor changes to sand and scattered cobbles. A trench running along the fault line runs the length of the cave.

Lava Bench Cave #2 is high and dry at all tide levels

LAVA BENCH CAVE #2

Middle Anacapa Island, California
Channel Islands National Park

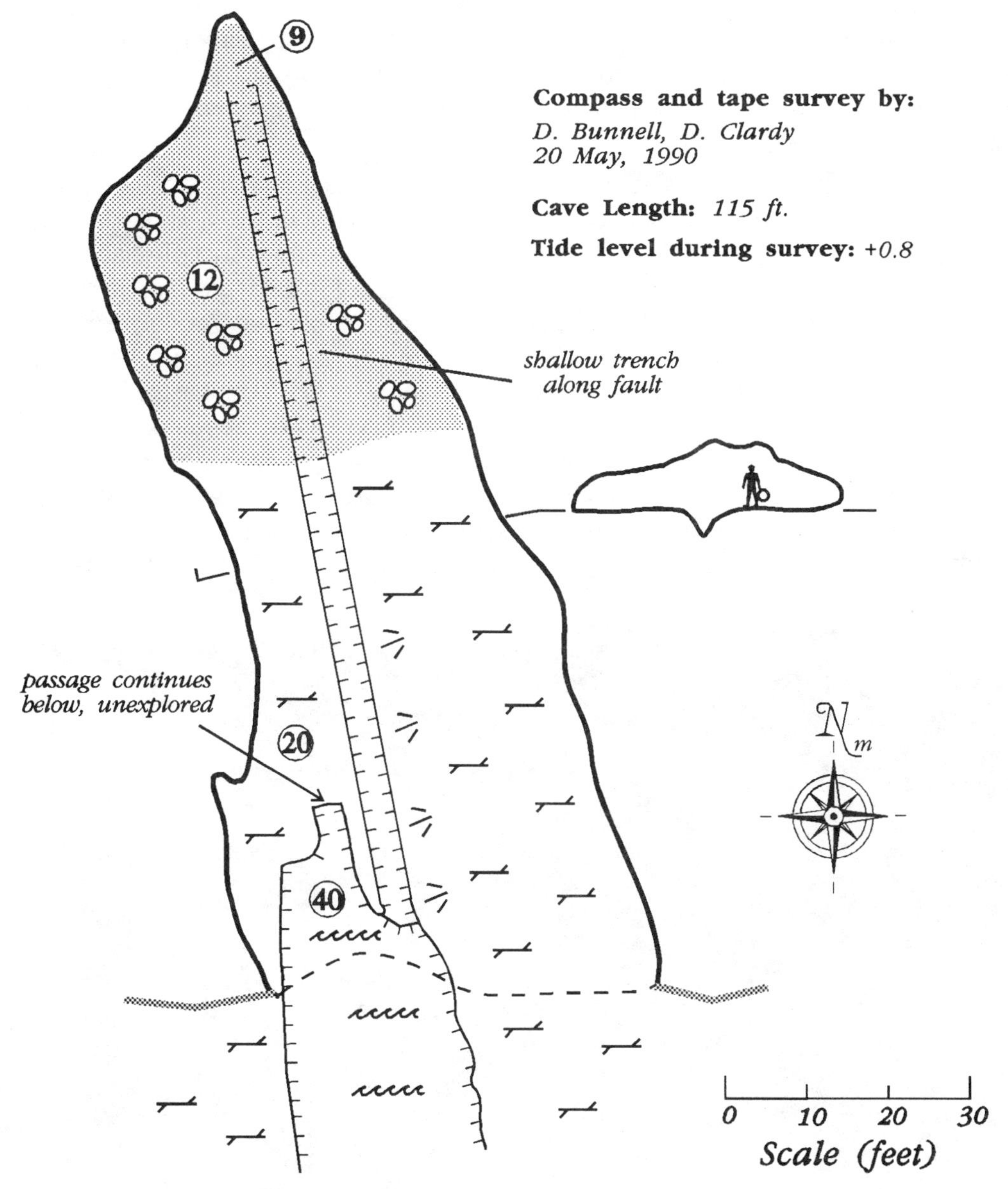

Compass and tape survey by:
D. Bunnell, D. Clardy
20 May, 1990

Cave Length: *115 ft.*

Tide level during survey: *+0.8*

Produced by Dave Bunnell

RESPIRING CHIMNEY CAVE - 40

Location: Westernmost of the caves on the lava bench west of East Fish Camp; a small cove lies just to the west.

Entrances: 2

Length: 448'

Conditions: The upper chamber can be safely explored with no special equipment. To safely explore the inner portions of the cave one should have: a helmet, a light (preferably one that can mount on the helmet), a wetsuit, and climbing skills. Tide level appears to be unimportant except that a negative tide may permit a "through-trip" between the entrances.

Description: This is one of the most unusual caves on the island, and the second largest of the south shore caves. The main cave at first appears to be simply a single dry chamber, like the nearby Lava Bench caves. In the rear, however, is a narrow crevice which drops 20' to a water passage. Air movement is evident at the top and was largely responsible for its discovery. It is a somewhat tricky climb, basically a chimney through crumbly rock. A handline might be useful for belay although there are no good anchor points. At the base, one direction heads back under the upper chamber and connects through a sump to the lower entrance. Continuing into the cave, one swims about 40' to a narrow climb-up. Next one walks on chockstones to a larger passage which cuts to the right. This leads 50' back to a constriction which opens into a larger, dry chamber, 40' wide and 15' high. No evidence of use by pinnipeds was observed. On the left wall of this chamber, a short crawling passage leads into a second chamber which contains a few stalactites.

The two entrances to Respiring Chimney Cave

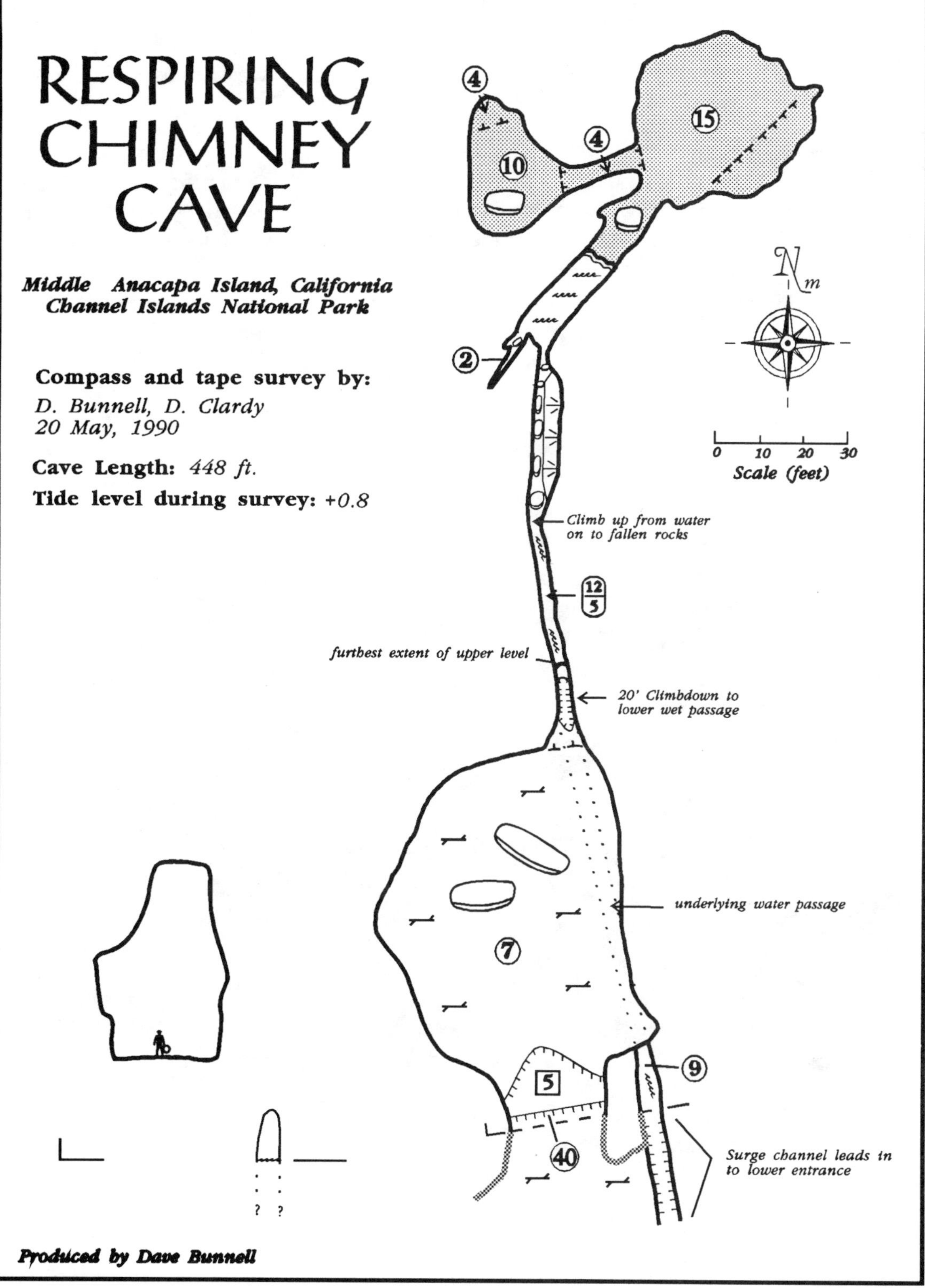
RESPIRING CHIMNEY CAVE
Middle Anacapa Island, California
Channel Islands National Park
Compass and tape survey by:
D. Bunnell, D. Clardy
20 May, 1990
Cave Length: 448 ft.
Tide level during survey: +0.8
Climb up from water on to fallen rocks
furthest extent of upper level
20' Climbdown to lower wet passage
underlying water passage
Surge channel leads in to lower entrance
0 10 20 30
Scale (feet)
Produced by Dave Bunnell

SUNBEAM CAVE - 41

Location: Easternmost of the cluster of caves located 2500' west of east Fish Camp

Length: 172.5'

Entrances: 2

Conditions: This cave may be explorable by dinghy or kayak at a moderate tide level, but at low tide would require wading through submerged rocks. A dinghy or kayak could be pulled up on the rocky benches near the entrance. The cave is fairly well illuminated throughout, but beware of urchins!

Description: Though only moderate in length, the cave's passage is somewhat wider than most and quite aesthetically pleasing. The cave has two entrances. The righthand entrance is low and opens only at a very low tide. The main entrance is to the left and is some 6' wide and 20' tall, opening immediately to a passage 20' wide and 16' high. This proceeds to a sloping cobble beach, where it widens to 30'. A third of the way back, two low passages are found on the left wall.

The two entrances to Sunbeam Cave

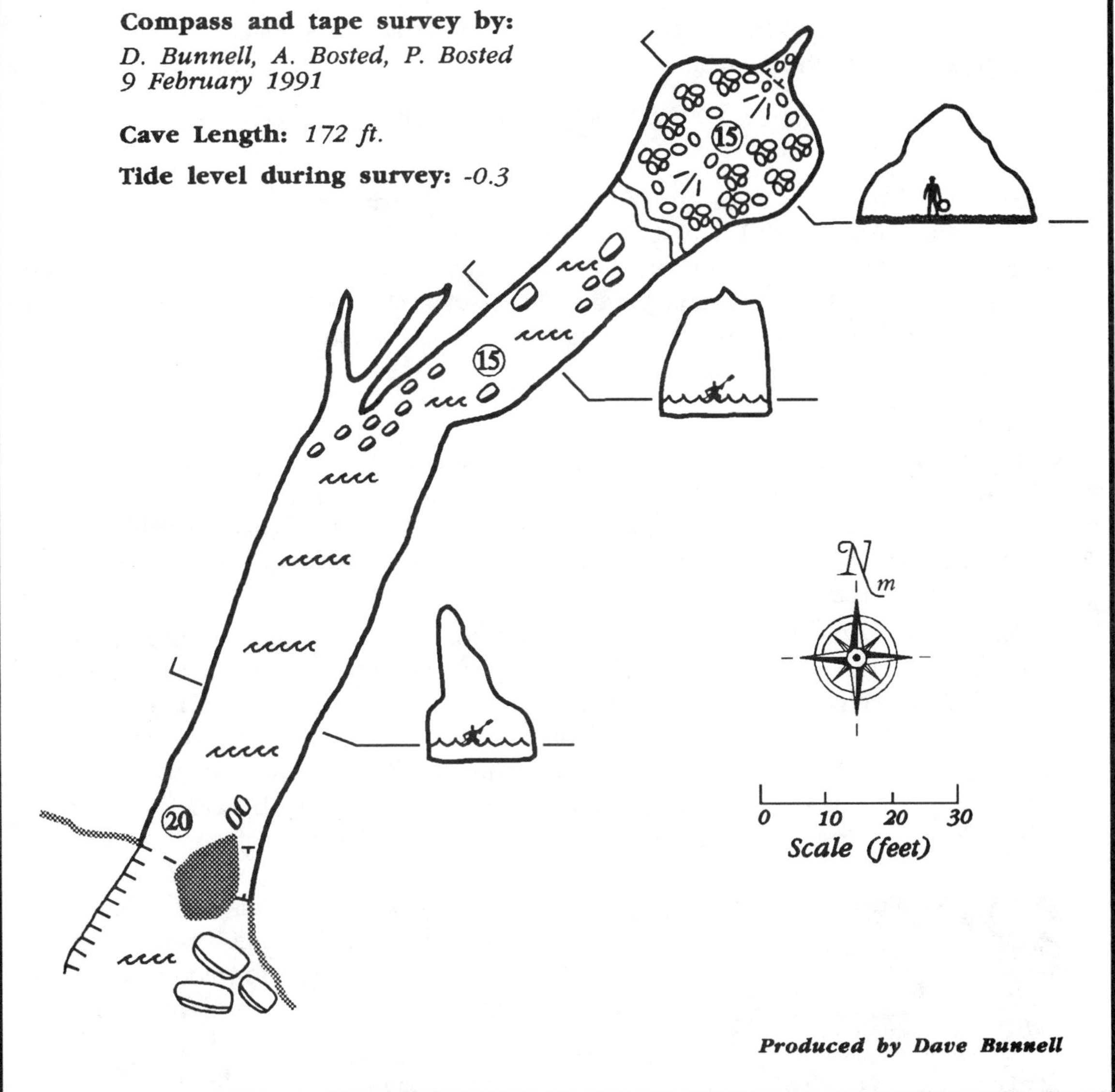
SUNBEAM CAVE
Middle Anacapa Island, California
Channel Islands National Park
Compass and tape survey by:
D. Bunnell, A. Bosted, P. Bosted
9 February 1991
Cave Length: 172 ft.
Tide level during survey: -0.3
15
15
20
Nm
0 10 20 30
Scale (feet)
Produced by Dave Bunnell

CRABBY CAVE - 42

Location: Approximately 2500' west of East Fish Camp. The entrance is somewhat hidden but is just to the right of Three Door Cave's more noticeable entrances.

Entrances: 1

Length: 149'

Conditions: The cave is most readily explored by landing nearby and wading in at low tide, as most of it is submerged at high tide. Bring a light, particularly to avoid urchins.

Description: This is a pleasant little cave with some nice tidepool life. Ceiling heights average 7 to 9' and the passage is 9 to 12' wide.

THREE DOOR CAVE - 43

Location: The cave's three entrances are the most prominent of the group of caves located approximately 2500' west of East Fish Camp. About 50' of cliff-face separate the entrances, which all face southward. About 75' west of the western entrance of 3-Door Cave is a prominent fault with an inverted "v"-shaped alcove and a fissure. We have never entered this apparent cave, which seemed intimidating even in a calm sea.

Entrances: 3

Length: 590'

Conditions: The cave is most readily explored by landing nearby and wading in at low tide. The connection through the "hidden door" is exposed only at lower minus tides. Most of the cave is of walking height. Bring a light, particularly to avoid urchins. Not a cave to explore when seas are rough, as there are numerous submerged rocks and some low spots in the rear.

Description: This impressive cave is formed from the intersection of three separate cave passages. The leftmost entrance leads through water-floored passage to a fork. The righthand fork leads into a low-ceiling area 4' high, and give access to the "hidden door", a 2' high crawlway on the right. The leftmost fork leads 60' to a cobble beach and also connects midway into the righthand fork. Entrances 2 and 3 lead to passages on converging faults. The two main passages each average some 15' wide and form a cobble-floored chamber 48' wide upon joining. A portion of wall which sparated the two passages before they joined is still standing but is now undercut, forming a very large pendant. Another portion of wall remains as a pillar.

The 3 entrances to Three Door Cave. Crabby Cave entrance not visible at this angle.

THREE DOOR & CRABBY CAVES

Middle Anacapa Island, California
Channel Islands National Park

Compass and tape survey by:

A. Bosted, P. Bosted, D. Bunnell
9 February 1991

D. Bunnell, D. Clardy, L. DiThomas, L. Schultz
24 November 1991 (connection of west entrance)

Tide level during survey: *-0.3*

Tide level during 2nd survey: *-0.5*

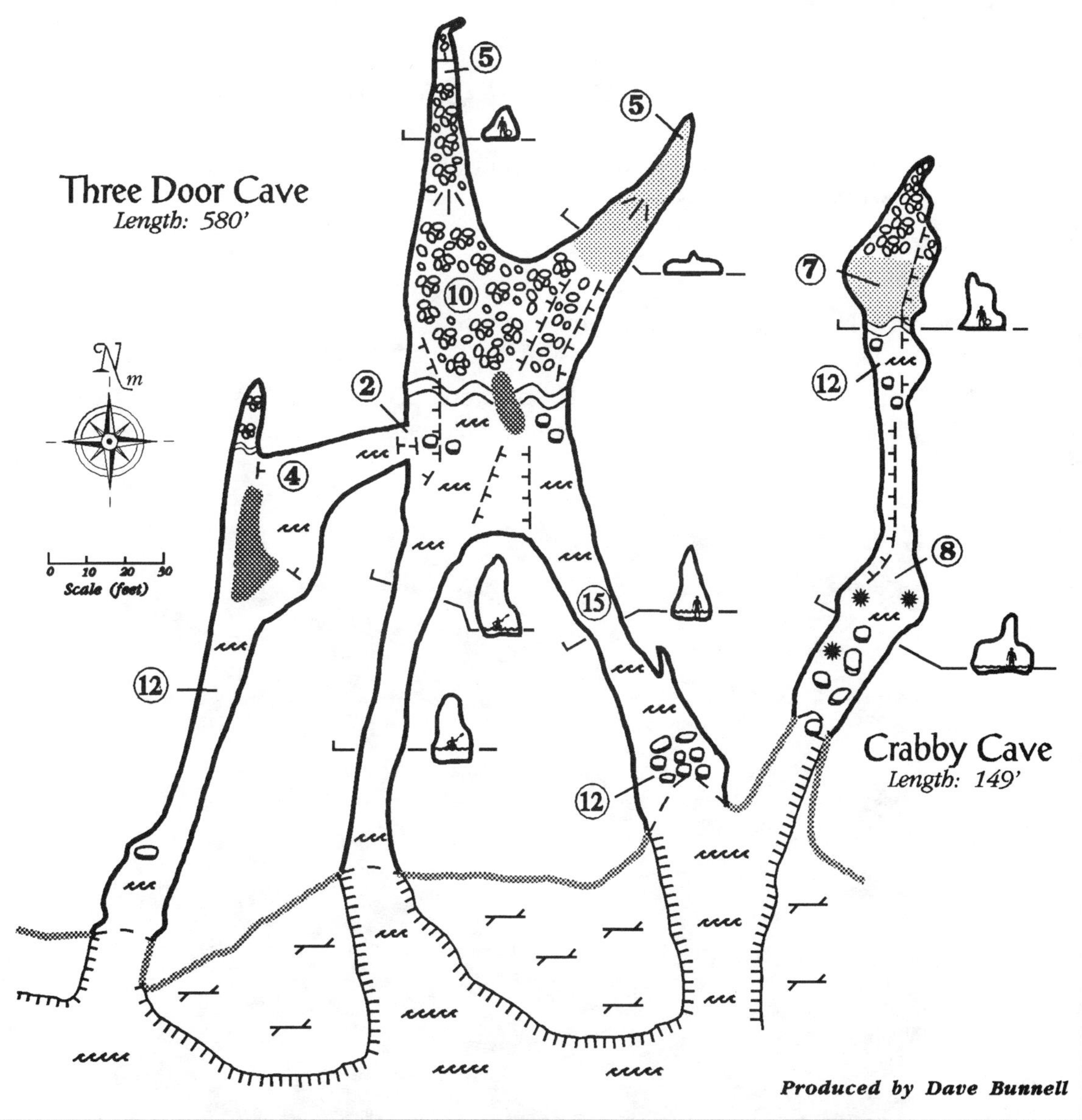

DOGLEG CAVE - 44

Location: On the east side of a small beach 4000' west of East Fish Camp (3000' east of the west tip of Middle Island).

Entrances: 1

Length: 132'

Conditions: Not explorable by dinghy. Land on the small beach and walk in at low tide or swim in at higher tides; a light is useful.

Description: The cave's entrance is 10' high and 9' wide; inside it averages some 10' high by 9' wide. Much of it is floored with water at low tide and probably most of it is submerged at high tide. The passage curves to the left out of direct view of the entrance, and ends in a small cobble-floored alcove.

SEALS' BEACH CAVE -45

Location: On the west side of a small beach 4000' west of East Fish Camp, or 3000' east of the west tip of Middle Island

Entrances: 1

Length: 84'

Conditions: Not explorable by dinghy. Land on the small beach and walk in at low tide; a light is useful.

Description: The cave's entrance is 10' high and 15' wide; inside it averages some 12' high by 5' wide. Much of it is floored with cobbles and most of it is submerged at high tide.

Dogleg Cave (far right) and Seal's Beach Cave (far left)

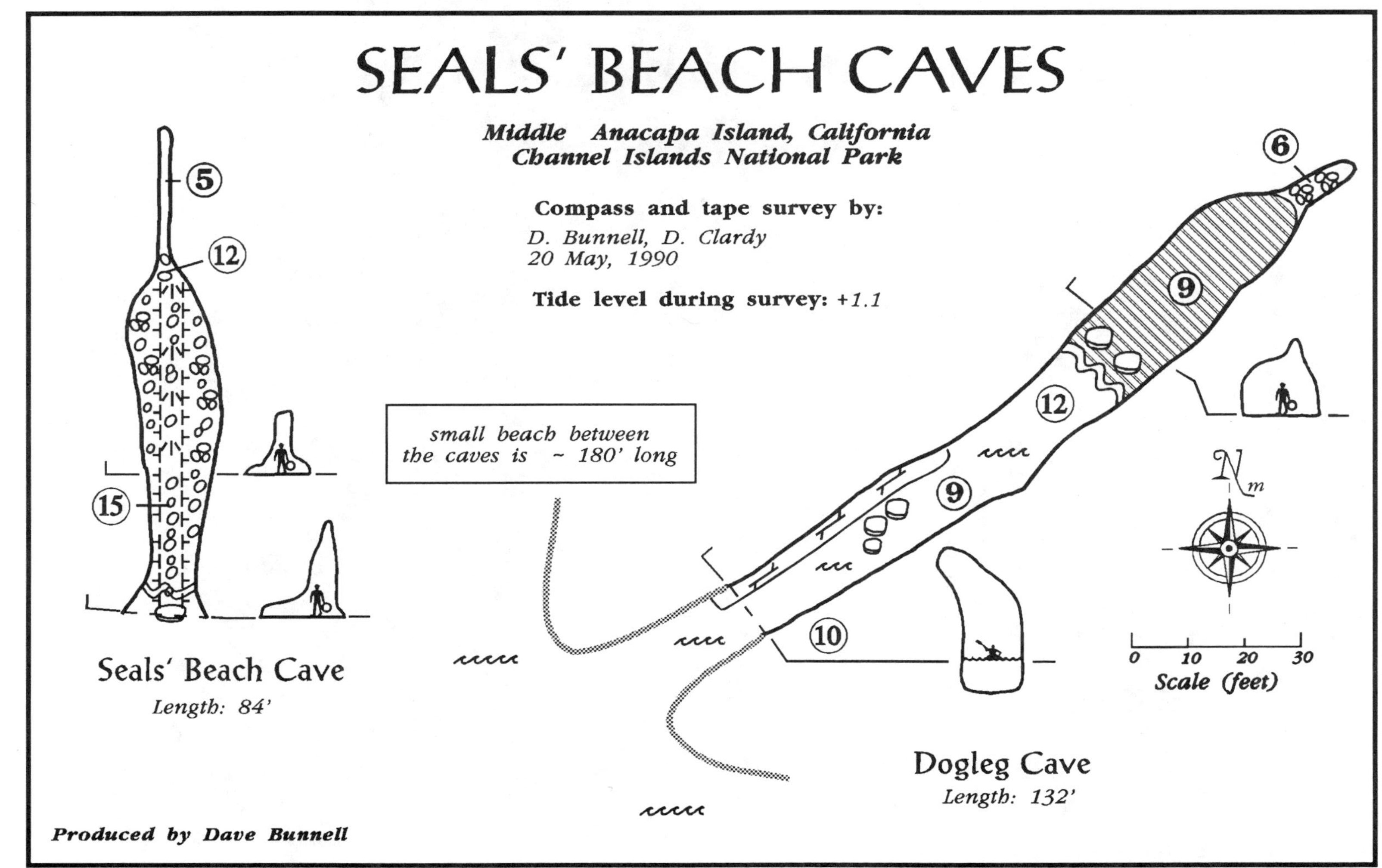
SEALS' BEACH CAVES
Middle Anacapa Island, California
Channel Islands National Park
Compass and tape survey by:
D. Bunnell, D. Clardy
20 May, 1990
Tide level during survey: +1.1
small beach between the caves is ~ 180' long
Seals' Beach Cave
Length: 84'
Dogleg Cave
Length: 132'
0 10 20 30
Scale (feet)
Produced by Dave Bunnell

DEEP PENETRATION CAVE - 46

Location: 2300' east of the west tip of Middle Island, on the edge of a small cove.

Length: 185'

Entrances: 1

Conditions: The cave faces directly into the west swell, so check prevailing conditions carefully. Even on a relatively calm day, some pretty rough sets came in to the cave during our survey. Optimally, one would explore this cave by snorkeling or swimming in. A kayak or dinghy could enter if conditions were unusually calm, probably at a higher tide level than during our survey since the surf would break further into the cave. A turn-around could be problematic however, depending on the length of the vessel. Check the map.

Description: The entrance is some 15' high and 7' wide. The cave is water-floored for the first 130' or so, with the surf crashing on a sandy beach in the rear. From the beach the main passage continues on some 40' as a sand-floored hands-and-knees crawl. This cave has one of the longer penetrations into the cliff-face on the south shore. This is not a deep-water cave, so of little interest to divers.

We approached Deep Penetration by landing kayaks on the beach just west of it and walking on the bedrock benches exposed at low tide.

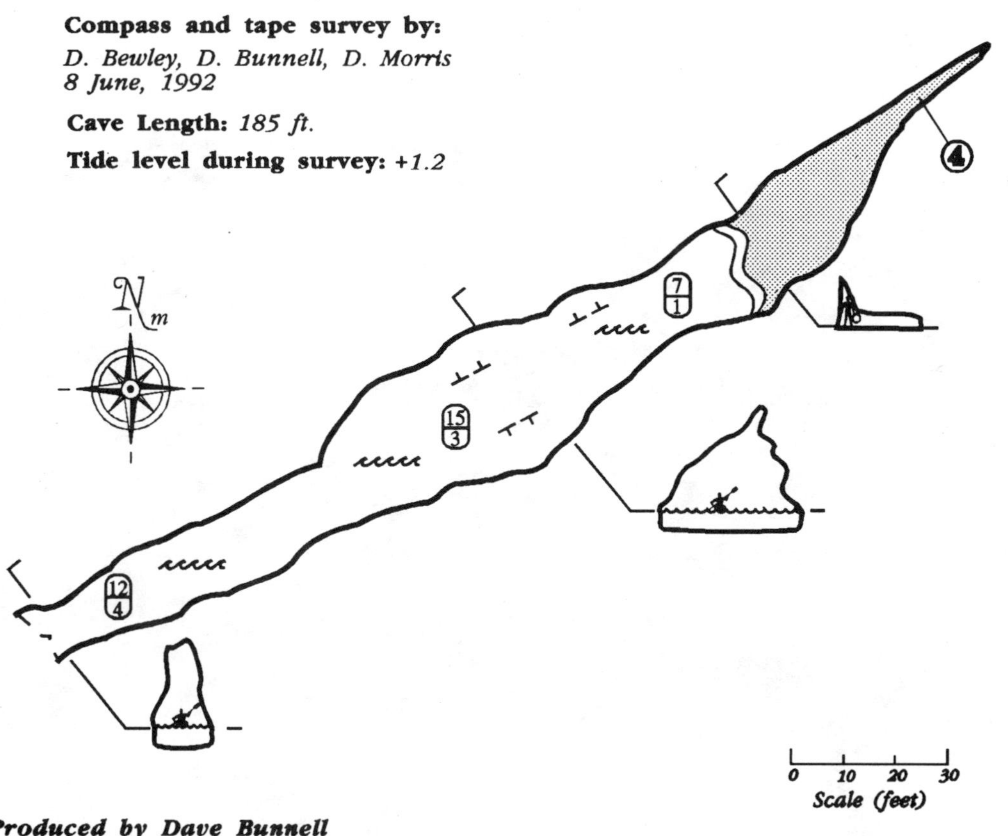
DEEP PENETRATION CAVE
Middle Anacapa Island, California
Channel Islands National Park
Compass and tape survey by:
D. Bewley, D. Bunnell, D. Morris
8 June, 1992
Cave Length: 185 ft.
Tide level during survey: +1.2
Nm
4
7
1
15
3
12
4
0 10 20 30
Scale (feet)
Produced by Dave Bunnell

COBBLE COVE CAVES - 47 & 48

Location: 2100' east of the west tip of Middle Island, in a small cove.

Lengths: Cave #1: 50'; Cave #2: 36'.

Entrances: 1 (for each cave)

Conditions: Both caves can be entered on foot at low tide, when they are dry. No lights are needed.

Description: These small caves are both entirely floored with cobbles. The larger cave has an entrance 12' high and 38' wide and is a single chamber. The smaller cave's entrance is 7' high and 12' wide, leading into a passage which quickly narrows into a tight fissure which has collected numerous flotsam.

The author in Cobble Cove Cave #1

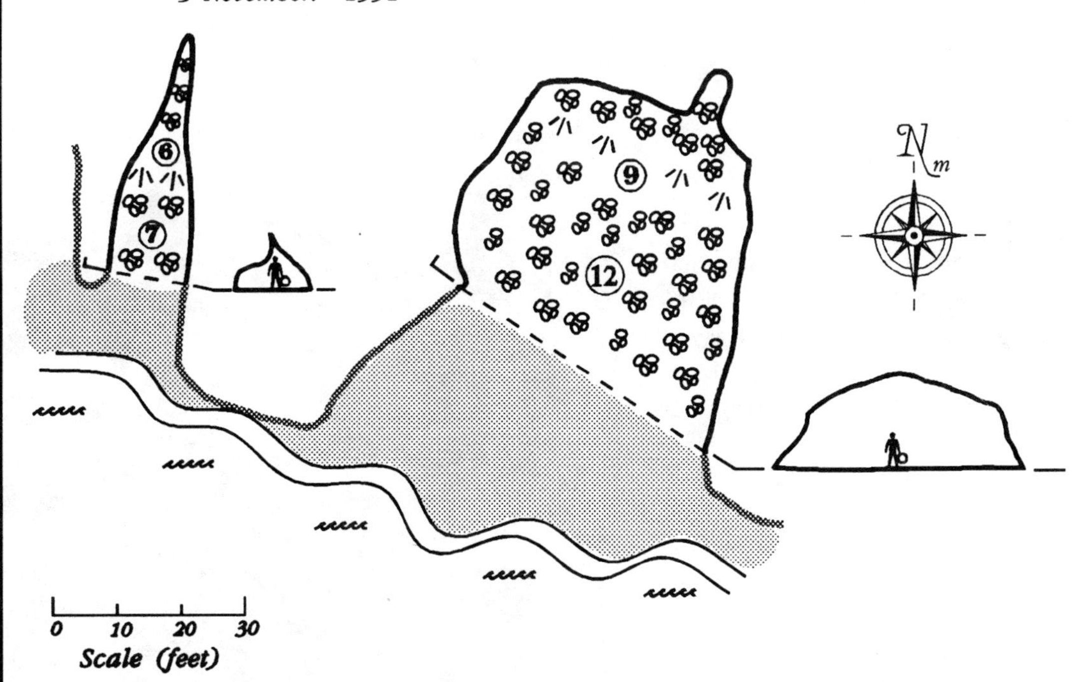
COBBLE COVE CAVES
Middle Anacapa Island, California
Channel Islands National Park
Compass and tape survey by:
D. Bunnell, E. Garza
5 November. 1991
Cave Lengths: 36 ft. and 50 ft.
Tide level during survey: -0.5
6
7
9
12
N m
0 10 20 30
Scale (feet)
Produced by Dave Bunnell

TIDAL LAGOON CAVE - 49

Location: 1800' east of the west tip of Middle Island.

Length: 58'

Entrances: 1

Conditions: This is a mellow little cave which empties out at low tide. No light is necessary to reach the back.

Description: The cave has a prominent entrance, 20' high and 12' wide, with an associated fault that runs to the top of the cliffs. A fallen rock blocks part of the entrance. At low tide, a 20' long tidepool, containing a few purple urchins, remained at the entrance. Beyond, the cave is floored with cobbles which slope up to a fissure at the end. There appears to be some development along an upper level (see photo) but this was not reached.

CLIFF CHASM - 50

Location: 1750' east of the west tip of Middle Island

Length: 36'

Entrances: 2

Conditions: A cave to visit when the tide is low and seas are calm.

Description: This cave follows a fissure paralleling the cliff face. The cleft itself is some 80' long but only 36' of it is roofed over. The west entrance is 10' wide and 6' high and leads into a somewhat wider chamber floored with colorful sponges and other tidepool life. A few feet beyond the cave narrows to 2-3' wide and opens through a second entrance into a 10' deep surge channel.

Tidal Lagoon Cave; Cliff Chasm is just to left, out of the picture

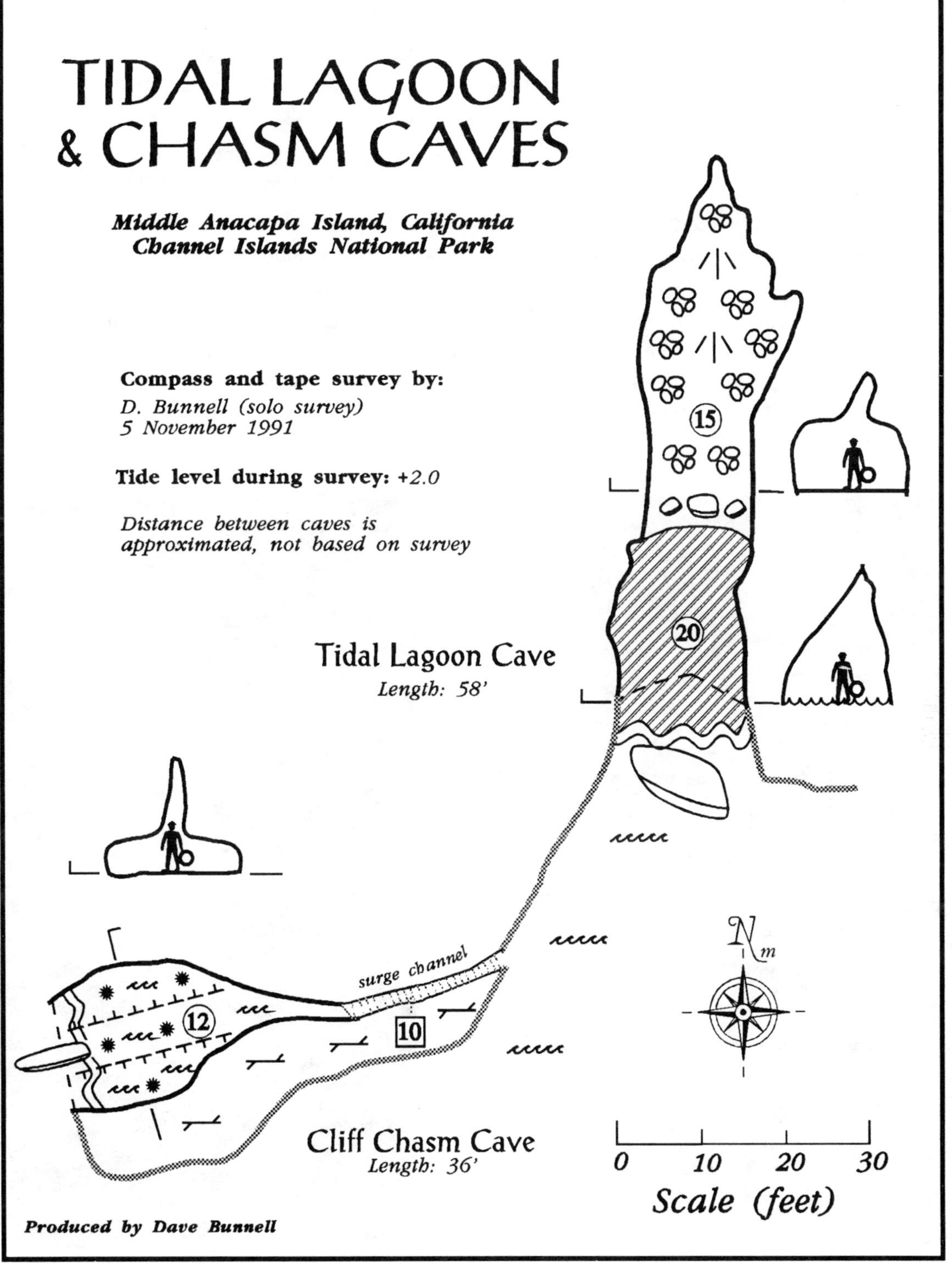

TIDAL LAGOON
& CHASM CAVES
Middle Anacapa Island, California
Channel Islands National Park
Compass and tape survey by:
D. Bunnell (solo survey)
5 November 1991
Tide level during survey: +2.0
Distance between caves is
approximated, not based on survey
Tidal Lagoon Cave
Length: 58'
15
20
surge channel
12
10
Cliff Chasm Cave
Length: 36'
0
10
20
30
Scale (feet)
Produced by Dave Bunnell

SANDY SURPRISE CAVE - 51

Location: 1000' east of the west tip of Middle Island, on the west edge of the westernmost cove on the south shore of the island. The prominent fault runs high up the cliff-face.

Length: 155'

Entrances: 1

Conditions: The cave can be entered on foot at low tide after pulling up on the beach nearby (but beware rocks). A light is needed to explore the side passage.

Description: The entrance is 7' wide and 25' high, and leads through fallen rocks to a cave floored entirely by sand. The main trend is 85' long, mostly walking-height.. About 2/3 of the way back, a dark side passage extends to the left for 70'. The side passage averages 5-6' high for about 50' then lowers to a crawlway about 2' high. At the junction the cave reaches its largest proportions, 15' high and 18' wide.

Sandy Surprise Cave is formed on a prominent fault

Looking out

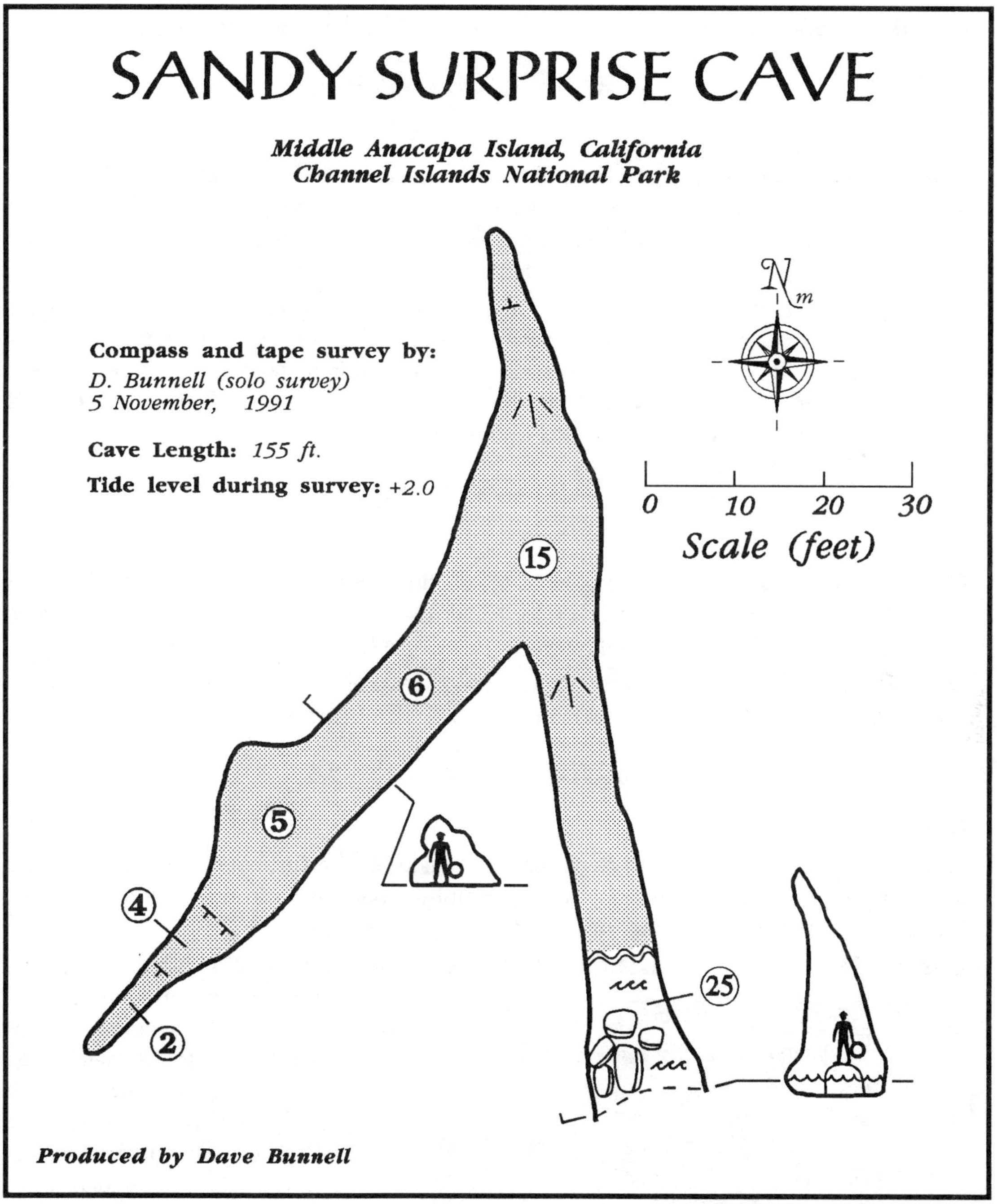

SANDY SURPRISE CAVE

Middle Anacapa Island, California
Channel Islands National Park

Compass and tape survey by:
D. Bunnell (solo survey)
5 November, 1991

Cave Length: *155 ft.*

Tide level during survey: *+2.0*

0 10 20 30
Scale (feet)

15

6

5

4

2

25

Produced by Dave Bunnell

THE GANG OF THREE

Location: 800' east of the western tip of Middle Island, on the south shore.

URCHINS' LAIR CAVE - 52

Length: 92'

Conditions: The cave is best explored at low tide, when one can wade through shallow water. Caution is required as there are numerous purple urchins. Rocks on the floor and at the entrance preclude use of kayaks or dinghies. Calm seas are needed, as a surge channel appears at low tide. A light is useful for viewing the copious tidepool life.

Description: This is a single-passage cave formed along two converging faults bearing 25° and 15°. The front portions of the cave are fairly shallow waters grading into tidepools amongst fallen rocks. Red and orange sponge are inter-mingled with urchins. In the rear is a sandy beach with cobbles strewn along the righthand wall. Most of the cave is walking-height except the rear portions of the beach.

PUD CAVE - 53

Length: 55'

Conditions: The cave is best explored at low tide, since there is little ceiling space at high tide. The cave must be explored on foot. Calm seas are needed, as there are numerous fallen rocks making footing difficult in a surge. A light is useful to reach the back.

Description: This is a standard single-passage cave along a fault bearing 15°. There is considerable fallen rock before one reaches a cobble beach in the rear.

PNEUMATIC CAVE - 54

Length: 124'

Conditions: The cave is best explored at low tide, as the surf zone reaches deep into the cave, even at low tide. A light is very useful for navigation in the rear of the cave. Exploration involves a combination of snorkeling and wading.

Description: This is a standard single-passage cave along a fault bearing 350°. This difference in orientation may account for its larger size in comparison to its neighbors, and indeed, the surf seemed rougher in this cave. It is almost all walking-height and water-floored, with submerged rocks to make the going more difficult.

The Gang of Three: from left to right, Pneumatic, Pud, and Urchin's Lair Caves

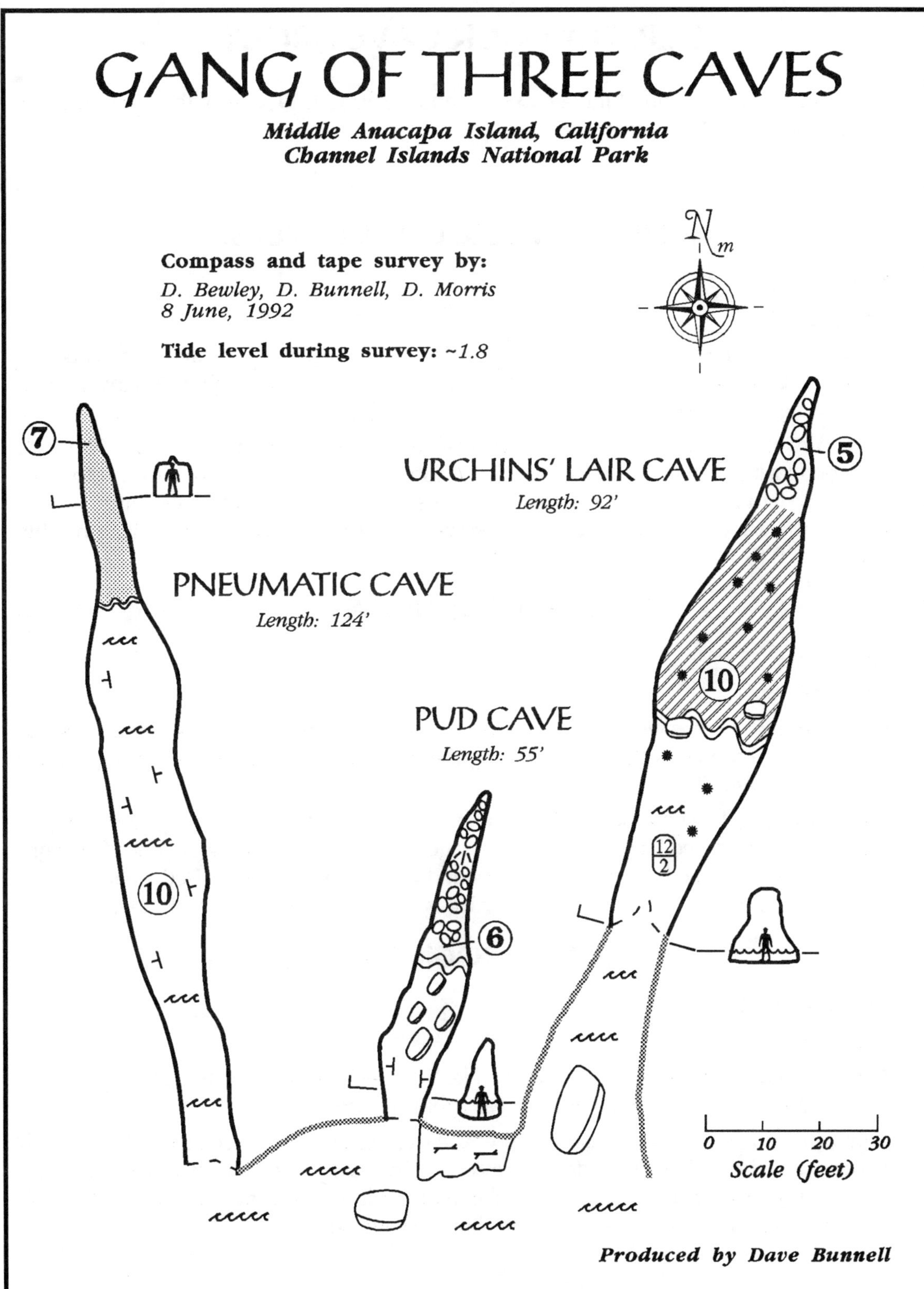
GANG OF THREE CAVES
Middle Anacapa Island, California
Channel Islands National Park
Compass and tape survey by:
D. Bewley, D. Bunnell, D. Morris
8 June, 1992
Tide level during survey: ~1.8
Nm
URCHINS' LAIR CAVE
Length: 92'
PNEUMATIC CAVE
Length: 124'
PUD CAVE
Length: 55'
7
5
10
10
6
12
2
0 10 20 30
Scale (feet)
Produced by Dave Bunnell

SLIPPERY ROCK CAVE GROUP

Location: This group of three caves lies about 250 to 400' east of the western tip of Middle Island, on the south shore.

RIPPLING REFLECTIONS CAVE - 55

Entrances: 1

Length: 117'

Conditions: The cave is best explored at low tide, since there is little ceiling space at high tide. Calm seas are needed, as a surge channel appears at low tide. A light is useful for viewing the copious tidepool life.

Description: This is a standard single-passage cave along a prominent fault. Although a second fault is visible just to the left of the entrance, the passage beneath it is too low and narrow to pursue. Red algae extends almost to the ceiling, and honeycomb worms encrust the walls below it. At low tide the water is waist to ankle-deep. Ceiling heights vary from 8 to 10'.

PURPLE TURMOIL CAVE - 56

Entrances: 1

Length: 85'

Conditions: A cave which is best visited at low tide and when seas are calm, since the surge channel amplifies the swell entering the cave. It was rough inside on a generally calm day as a result.

Description: The entrance is 12' wide and perhaps 15' tall. The passage is developed along a slightly angled fault. Inside the passage widens and the ceiling reaches 20' high. 45' past the dripline the passage narrows and extends another 40'. Red algae reaches almost to the ceiling, indicating that the cave is largely submerged at high tide.

SLIPPERY ROCK CAVE -57

Entrances: 1

Length: 73'

Conditions: This is a cave to visit at low tide. Entry will have to be on foot after swimming or kayaking to the rock bench nearby. No light is needed.

Description: This is a small but pretty cave, containing an isolated tidepool with numerous urchins. The cave ends in a crevice which has trapped numerous floats, buouys, etc. A long surge channel containing fallen rock leads into the cave and one must climb over these to enter. The rocks are quite slippery with algae at low tide.

Purple Turmoil (left) and Rippling Reflections (right) caves

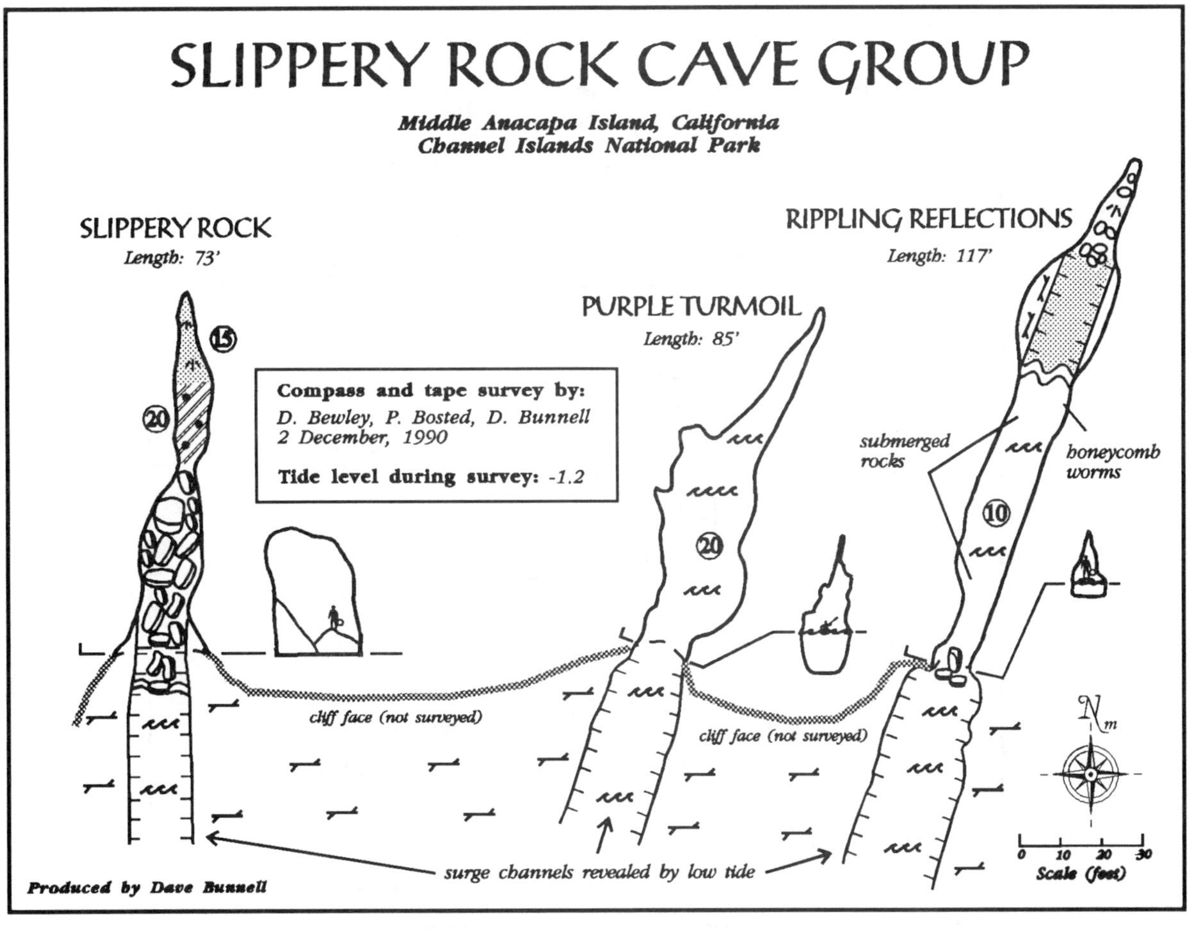

SLIPPERY ROCK CAVE GROUP

Middle Anacapa Island, California
Channel Islands National Park

SLIPPERY ROCK
Length: 73'

PURPLE TURMOIL
Length: 85'

RIPPLING REFLECTIONS
Length: 117'

Compass and tape survey by:
D. Bewley, P. Bosted, D. Bunnell
2 December, 1990

Tide level during survey: *-1.2*

15
20
20
10

submerged rocks
honeycomb worms
cliff face (not surveyed)
cliff face (not surveyed)
surge channels revealed by low tide

N_m

0 10 20 30
Scale (feet)

Produced by Dave Bunnell

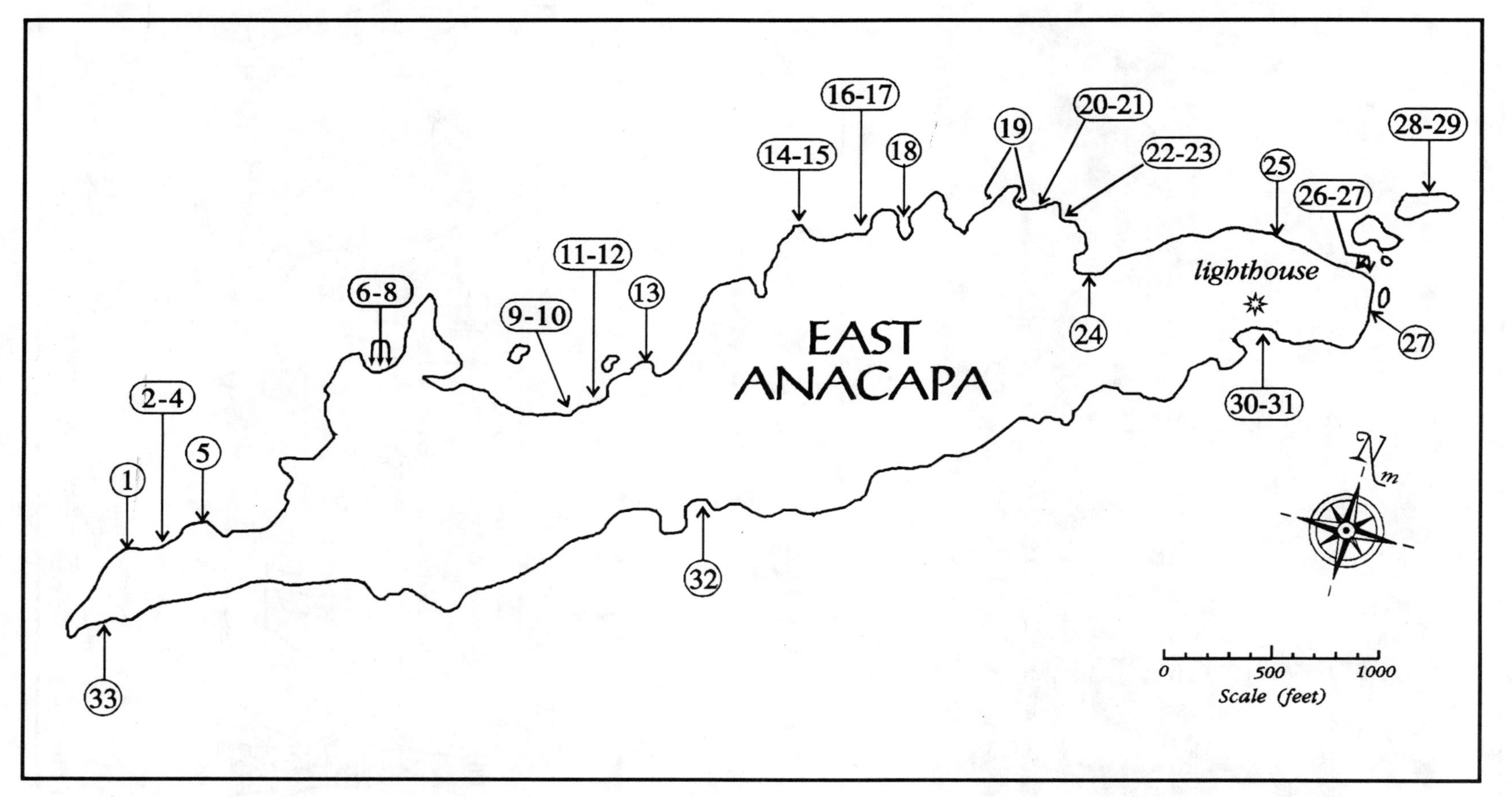
EAST
ANACAPA
lighthouse
1
2-4
5
6-8
9-10
11-12
13
14-15
16-17
18
19
20-21
22-23
24
25
26-27
27
28-29
30-31
32
33
N
m
0
500
1000
Scale (feet)

EAST ISLAND CAVES

NORTH SHORE

1 WEST OF EAST CAVE (125')
2 DINKY CAVE (86')
3 BIG COBBLE CAVE (210')
4 HIDDEN CAVE (47')
5 STARFISH CAVE (420')
6 CATHEDRAL CAVE (790')
7 HAPPY LOBSTER CAVE (114')
8 CATHEDRAL ARCH CAVE (144')
9 EL PEQUEÑO (52')
10 SEAL'S REFUGE CAVE (84')
11 MURKY WATER CAVE (66')
12 DEAD SEAL CAVE (328')
13 COLLAPSING CAVE (354')
14 CLAUSTROPHOBIC BEACH CAVE (162')
15 TIGHT END CAVE (110')
16 THE CATACOMBS (808')
17 CATACOMBS ARCH (58')
18 DEBRIS CAVE (232')
19 HALF GONE CAVE (462')
20 ROCKFALL CAVE (105')
21 OUTSIDE TUNNEL CAVE (353')
22 RECT. REFLECTIONS CAVE (154')
23 PENDANT CAVE (43.5')
23a. {cave feature - blowhole}.
24 LANDING COVE CAVE (148')
25 LA GRIETA (381')
26 EAST END CAVE (46')
27 EAST END TUNNEL (128')
28 ARCH ROCK CAVE WEST (75')
29 ARCH ROCK CAVE EAST (48')

SOUTH SHORE:

30 FEARFUL FISSURE (157')
31 LIGHTHOUSE CAVE (218')
32 LOST GLASSES CAVE (72')
33 TIP CAVE (41')

WEST OF EAST CAVE - 1

Location: Westernmost cave on East Island, north shore.

Entrances: 2

Length: 125'

Conditions: A cave best explored at low tide. A light is useful to negotiate submerged rocks and urchins. Not suitable for exploration by dinghy.

Description: A nice little cave with colorful tidepool llife and two entrances which lead into a 70' long chamber ending in a cobble beach.

The western end of East Island. One of the entrances to West of East Cave is just to the left of the surf line , or about a kayak's length to the right of the kayaker.

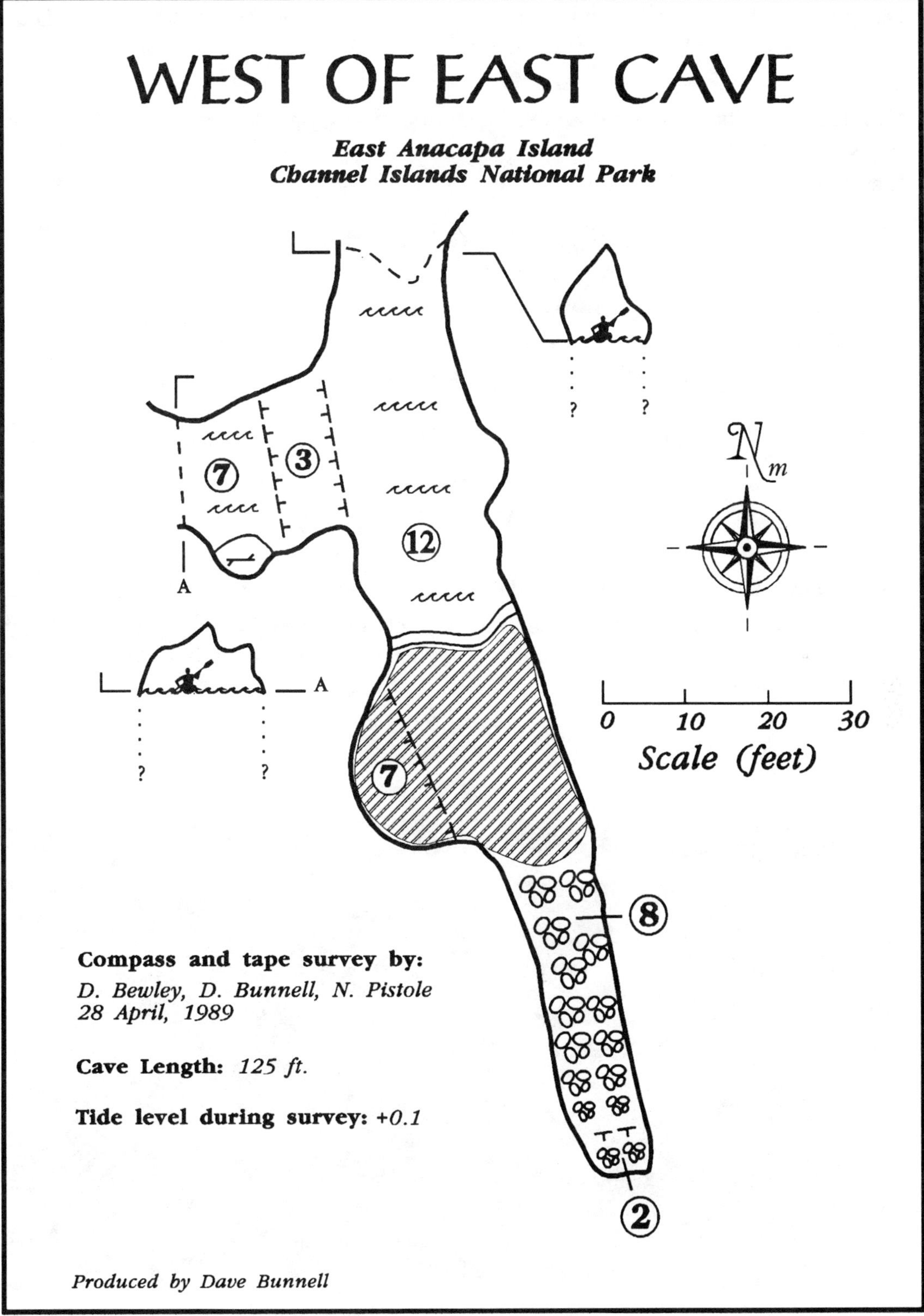
WEST OF EAST CAVE
East Anacapa Island
Channel Islands National Park
7
3
12
7
8
2
A
A
?
?
?
?
N
m
0
10
20
30
Scale (feet)
Compass and tape survey by:
D. Bewley, D. Bunnell, N. Pistole
28 April, 1989
Cave Length: 125 ft.
Tide level during survey: +0.1
Produced by Dave Bunnell

DINKY CAVE - 2

Location: Caves 2-4 are located in close proximity to each other, about 500' from the west tip of east island. The cliff here is heavily coated with white bird droppings. Hidden Cave was so named because it is difficult to spot until you are right in front of it.

Entrances: 1 **Length:** 86'

Conditions: The cave is explorable by dinghy unless the tide is very low. At low tide the water is knee-to-waist deep.

Description: The cave is a small room leading into a cobble-floored crevice.

BIG COBBLE CAVE - 3

Entrances: 1 **Length:** 210'

Conditions: Explore this cave on foot if seas are calm enough to land a dinghy on its cobble beaches. Keep right for a higher ceiling and watch out for submerged rocks. A light is necessary to explore the side passages.

Description: The cave is primarily a chamber 60' wide and somewhat over 100' deep, with three cobble-filled side passages.

HIDDEN CAVE - 4

Entrances: 1 **Length:** 47'

Conditions: Swim or tube to the entrance, or climb in from the rocks above it. A light is useful for the rear portions. The channel leading into the cave tends to be quite surgey. Explore at low tide.

Description: A single-fissure cave whose most interesting feature is its location and some nice tidepool life.

Big Cobble (left) and Dinky Cave entrances

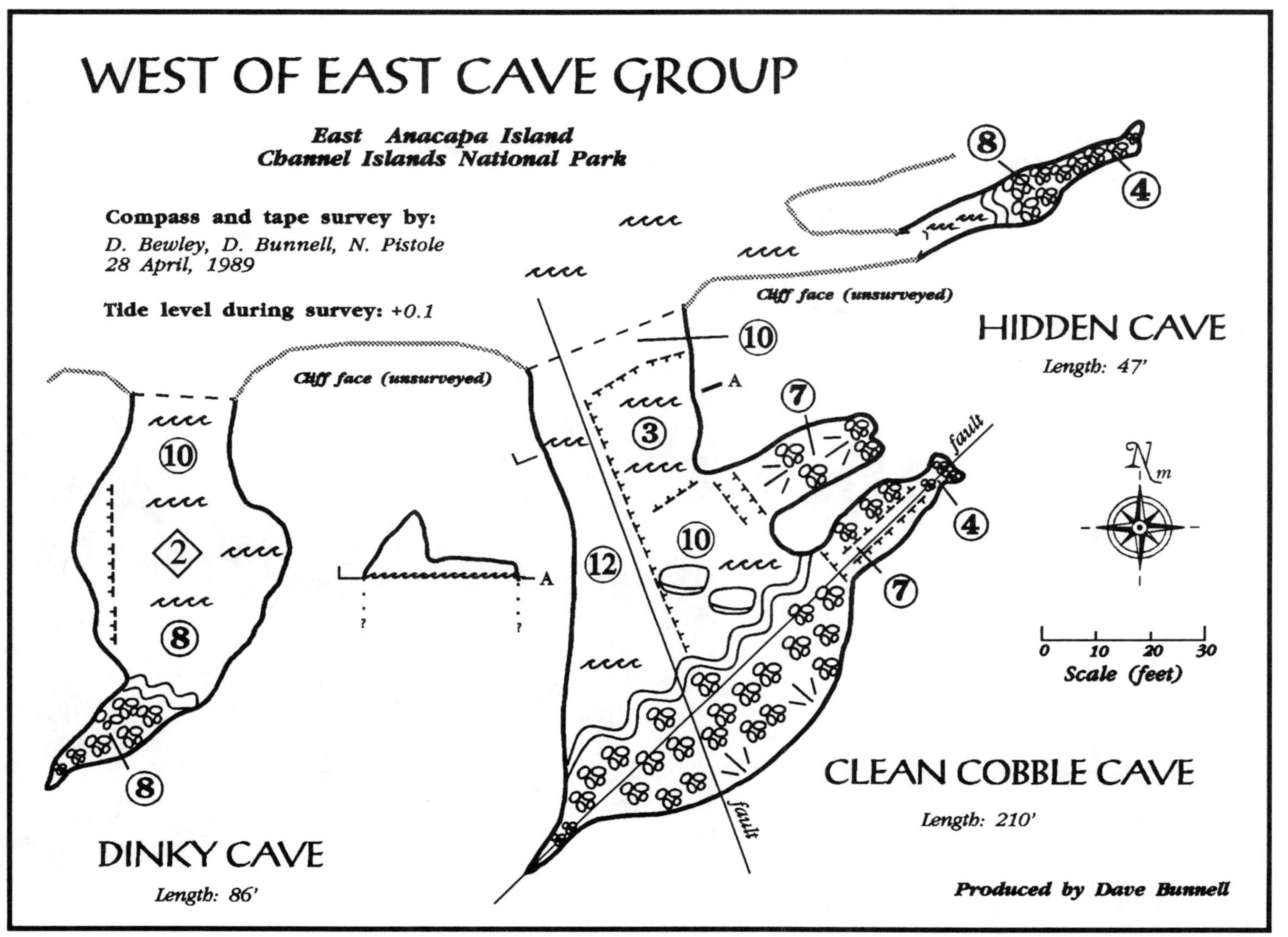
WEST OF EAST CAVE GROUP
East Anacapa Island
Channel Islands National Park
Compass and tape survey by:
D. Bewley, D. Bunnell, N. Pistole
28 April, 1989
Tide level during survey: +0.1
Cliff face (unsurveyed)
Cliff face (unsurveyed)
HIDDEN CAVE
Length: 47'
fault
fault
0 10 20 30
Scale (feet)
CLEAN COBBLE CAVE
Length: 210'
DINKY CAVE
Length: 86'
Produced by Dave Bunnell

STARFISH CAVE - 5

Location: Two of the three entrances are found in the west wall of the first significant cove encountered as one heads east from the western tip of East Island. The third and largest entrance is around the corner to the west of this cove.

Entrances: 3

Length: 420'

Conditions: A dinghy might be taken into the main, or western entrance, if seas are calm and the tide is moderate to high. Most of the cave must be explored on foot, however, and it is thus better explored by landing on the beach and entering the cave via the "windy link" passage at low tide, which still may require some wading. A light is needed to negotiate the floor and to avoid urchins.

Description: This is one of the nicest of the East island caves. It consists of a main chamber some 180' long, 7 to 15' high, floored with cobbles and rounded bedrock basins containing tidepools. There are three side passages. On the right side is a 45' long dead-end fissure passage. On the left side are two passages leading to the other entrances. The first of these is about 60' long, wet and low, and traversable only if the tide is low and seas are calm. The second is about 90' long and walking-height, with a key-shaped profile. Wind flows through the passage,hence its name.

Phil Darling in the Windy Link Passage of Starfish Cave

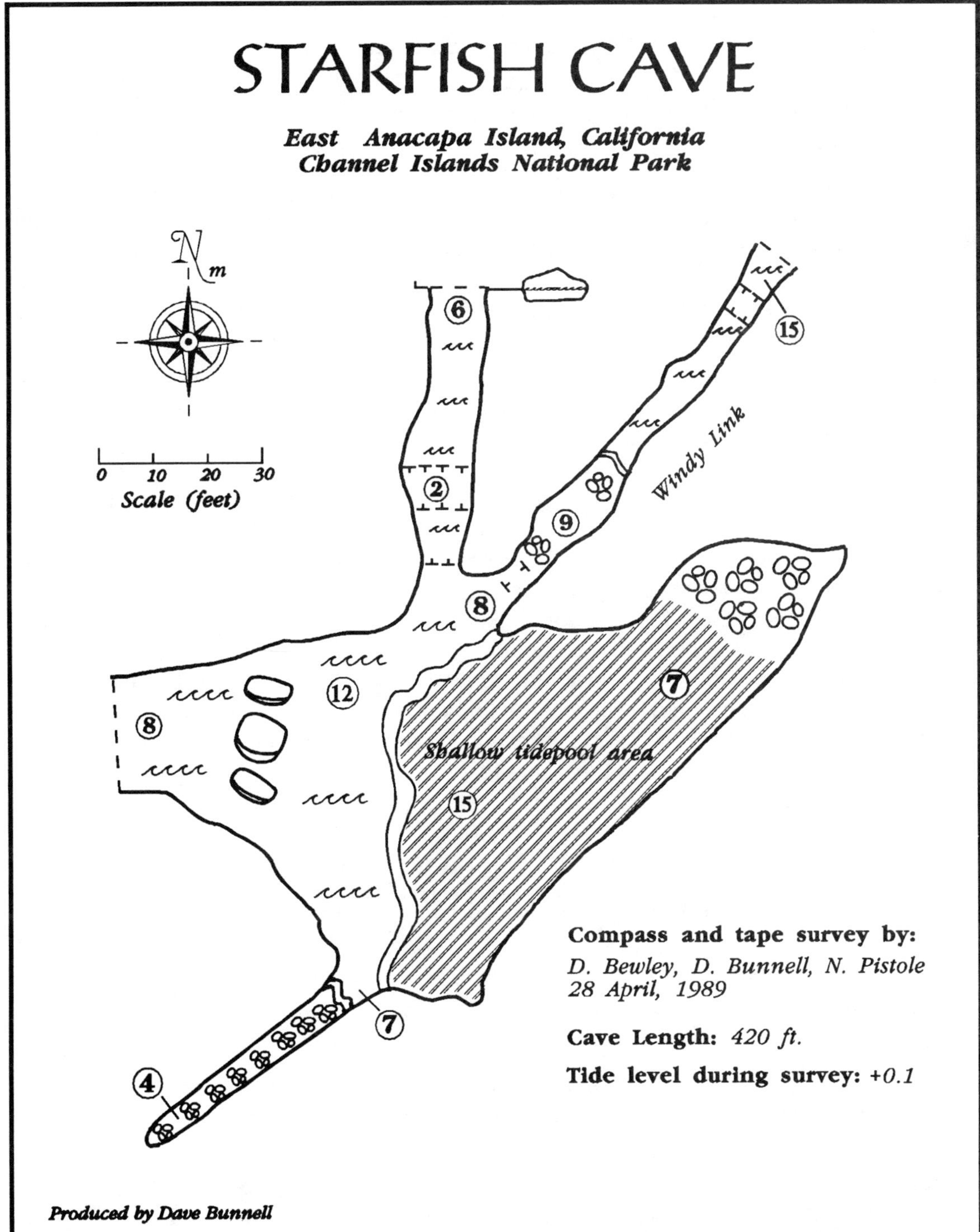
STARFISH CAVE
East Anacapa Island, California
Channel Islands National Park
m
0 10 20 30
Scale (feet)
6
2
8
9
Windy Link
15
12
8
7
Shallow tidepool area
15
7
4
Compass and tape survey by:
D. Bewley, D. Bunnell, N. Pistole
28 April, 1989
Cave Length: 420 ft.
Tide level during survey: +0.1
Produced by Dave Bunnell

CATHEDRAL CAVE - 6

Location: Just west of Cathedral Cove on East Anacapa. Indicated on Park Service maps.

Entrances: 5; Two of the entrances are submerged except at low tide.

Length: 790'

Conditions: The cave is best explored with a dinghy or kayak, at a tide level above 0.0 or so. With a lower tide submerged rocks may present a hazard, particularly in surgey conditions. Such conditions prevail here all too frequently, since the numerous entrances leave the cave open to most prevailing swell conditions. Ceiling heights are a comfortable 15 to 25' high so exploration should be possible at even the highest of tides. The cave is well illuminated by the numerous entrances, so lights are unnecessary.

Description: Cathedral is one of the largest and most spectacular caves on the island. Five entrances lead into a large "L" - shaped chamber. On the east side are two large entrances separated by a pillar, which give access to a large chamber. To the left the room trends southeast and ends in two dimly-lit, sloping-cobble beaches. Midway, a side passage gives access at low tide through a low point to another eastern entrance. If one goes straight (rather than turning into the chamber), the passage trends westward and is split by another pillar, beyond which the passage continues to two more entrances. The righthand entrance is low and may seal up at higher tides, admitting light underwater and giving a blue-green glow to the water. The main westward entrance emerges into the next cove. Just outside this entrance another passage to the left heads south for 135', averaging eight feet wide and seven to ten feet high, and is largely water-floored. Cathedral Cave is particularly beautiful at lower tides when numerous multi-colored sponges are exposed. In calm conditions it might be a good spot for snorkeling. Water depths in the cave seem to be ten feet or less throughout.

HAPPY LOBSTER CAVE - 7

Location: In-between Cathedral and Cathedral Arch Caves

Entrances: 1

Length: 114'

Conditions: Best explored at low tide in calm seas with snorkeling gear, as the passage is somewhat narrow for dinghies. A light is useful, especially for viewing the numerous lobster (remember, this is a protected habitat: look but don't touch!)

Description: This cave is a single water-floored passage averaging eight feet high and ending in a small cobble beach. It was named for the numerous lobster seen on the walls and practically leaping on us as we surveyed.

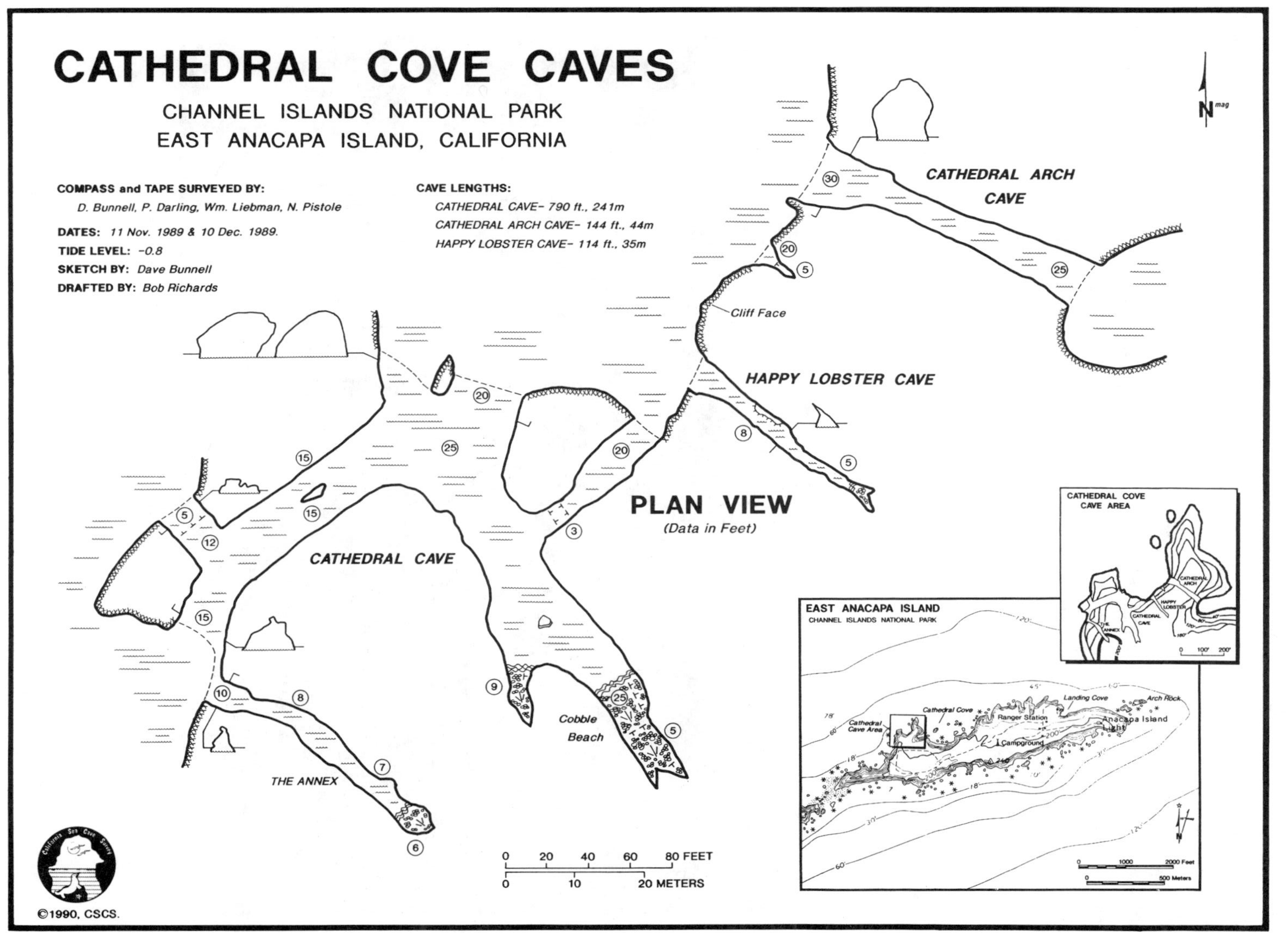
CATHEDRAL COVE CAVES
CHANNEL ISLANDS NATIONAL PARK
EAST ANACAPA ISLAND, CALIFORNIA
COMPASS and TAPE SURVEYED BY:
D. Bunnell, P. Darling, Wm. Liebman, N. Pistole
DATES: 11 Nov. 1989 & 10 Dec. 1989.
TIDE LEVEL: -0.8
SKETCH BY: Dave Bunnell
DRAFTED BY: Bob Richards
CAVE LENGTHS:
CATHEDRAL CAVE- 790 ft., 241m
CATHEDRAL ARCH CAVE- 144 ft., 44m
HAPPY LOBSTER CAVE- 114 ft., 35m
N mag
CATHEDRAL ARCH CAVE
Cliff Face
HAPPY LOBSTER CAVE
PLAN VIEW
(Data in Feet)
CATHEDRAL CAVE
Cobble Beach
THE ANNEX
0 20 40 60 80 FEET
0 10 20 METERS
CATHEDRAL COVE CAVE AREA
EAST ANACAPA ISLAND
CHANNEL ISLANDS NATIONAL PARK
Cathedral Cove
Landing Cove
Arch Rock
Ranger Station
Anacapa Island Light
Cathedral Cave Area
Campground
©1990, CSCS.

CATHEDRAL ARCH - 8

Location: Connects Cathderal Cove and the small cove just to the west, which contains three of the five entrances to Cathedral Cave.

Entrances: 2

Length: 144'

Conditions: Makes a nice through-trip in a kayak or dinghy if surge is minimal, which is more likely at a moderate or high tide. No lights are needed.

Description: An impressive tunnel averaging some 20' high and 10' wide connecting the cobble beach at Cathedral Cove with a small cove to the west. The beach entrance isn't visible except from the beach.

Kayaker emerges from a "through-trip" of the arch

Nancy Pistole makes the through-trip by kayak

The main entrance to Cathedral on the east side of the point

EL PEQUEÑO CAVE - 9

Location: In the middle of Cathedral Cove

Entrances: 1

Length: 52'

Conditions: This small cave is best explored at low tide and with a light, by swimming or tubing in.

Description: The entrance is 6' wide and 8' high, and leads into a 30' wide, 5' high cobble-floored room.

SEAL'S REFUGE CAVE - 9

Location: In the middle of Cathedral Cove

Entrances: 1

Length: 84'

Conditions: The cave is probably explorable at any tide level, and a light is needed to view the inner chamber. Don't visit when seals are hauled out inside.

Description: The cave has a broad arching entrance, 45' wide and 30' high. Some 20' from the dripline the ceiling lowers to 3 to 5' high and then opens into a dark inner chamber 10' high and 30' wide. This chamber is noticeably warmer and more humid than the outside. The air here is fouled by the odor of seal urine and is a most unpleasant place to linger. There is a small stalagmite 2" long in the rear of the cave. The cave is home to California sea lions on a fairly regular basis, but only three were noted at the time of our survey in July 1991.

Seal's Refuge Cave on left, El Pequeño Cave on right

SEAL'S REFUGE & EL PEQUEÑO CAVES

East Anacapa Island, California
Channel Islands National Park

Compass and tape survey by:
D. Bewley, D. Bunnell, R. Niemi
13 July, 1991

Tide level during survey: *+1.7*

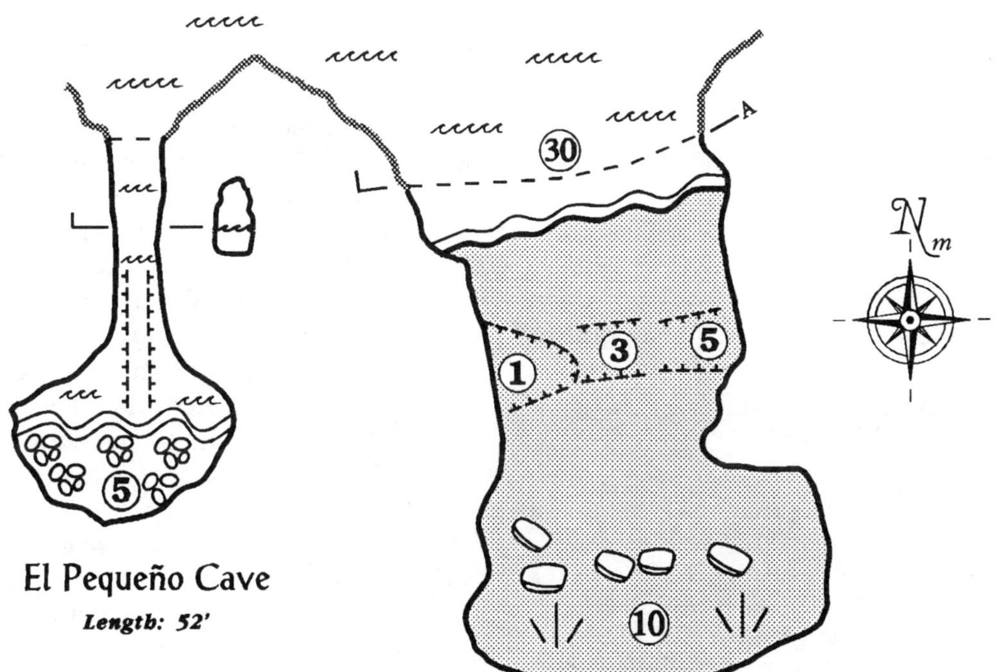

Seal's Refuge
Length: 84'

0 10 20 30 40 50
Scale (feet)

Produced by Dave Bunnell

MURKY WATER CAVE - 11

Location: In the east end of Cathedral Cove

Entrances: 1

Length: 66'

Conditions: One must swim into this cave, which doesn't require a particularly low tide. A light would be useful in the rear.

Description: The cave is a single chamber some 15' wide, 66' long, and about 8' high. At low tide there is a small cobble beach in the rear.

DEAD SEAL CAVE - 12

Location: East end of Cathedral Cove

Entrances: 2

Length: 328'

Conditions: This cave is explorable by dinghy or kayak, although low tides expose numerous submerged rocks and render this problematic. There is enough of a cobble beach in the rear at a moderate low tide to pull out a dinghy. The through-trip is mostly a walk-through at low tide and a swim-through at high tide. Lights aren't necessary.

Description: This cave has two entrances. The main entrance is 50' wide and 60' high and leads into a water-floored chamber. At the rear is a broad cobble beach with a large breakdown pile on the left. This rockpile is contiguous with that in nearby Collapsing Cave (#13), but we were unable to find any physical connection. A dead seal was lying on the rocks during our survey. On the righthand wall of the main chamber is an archway leading into two other passages. The shorter passage on the left is 50' long into a small breakdown-floored room. The righthand passage leads into a 90' long walking-height tunnel to another entrance. Numerous urchins on the floor of this passage require careful movement at low tide, and it is more easily traversed at a higher tide that permits swimming over the urchins.

Dead Seal Cave with a live seal in front

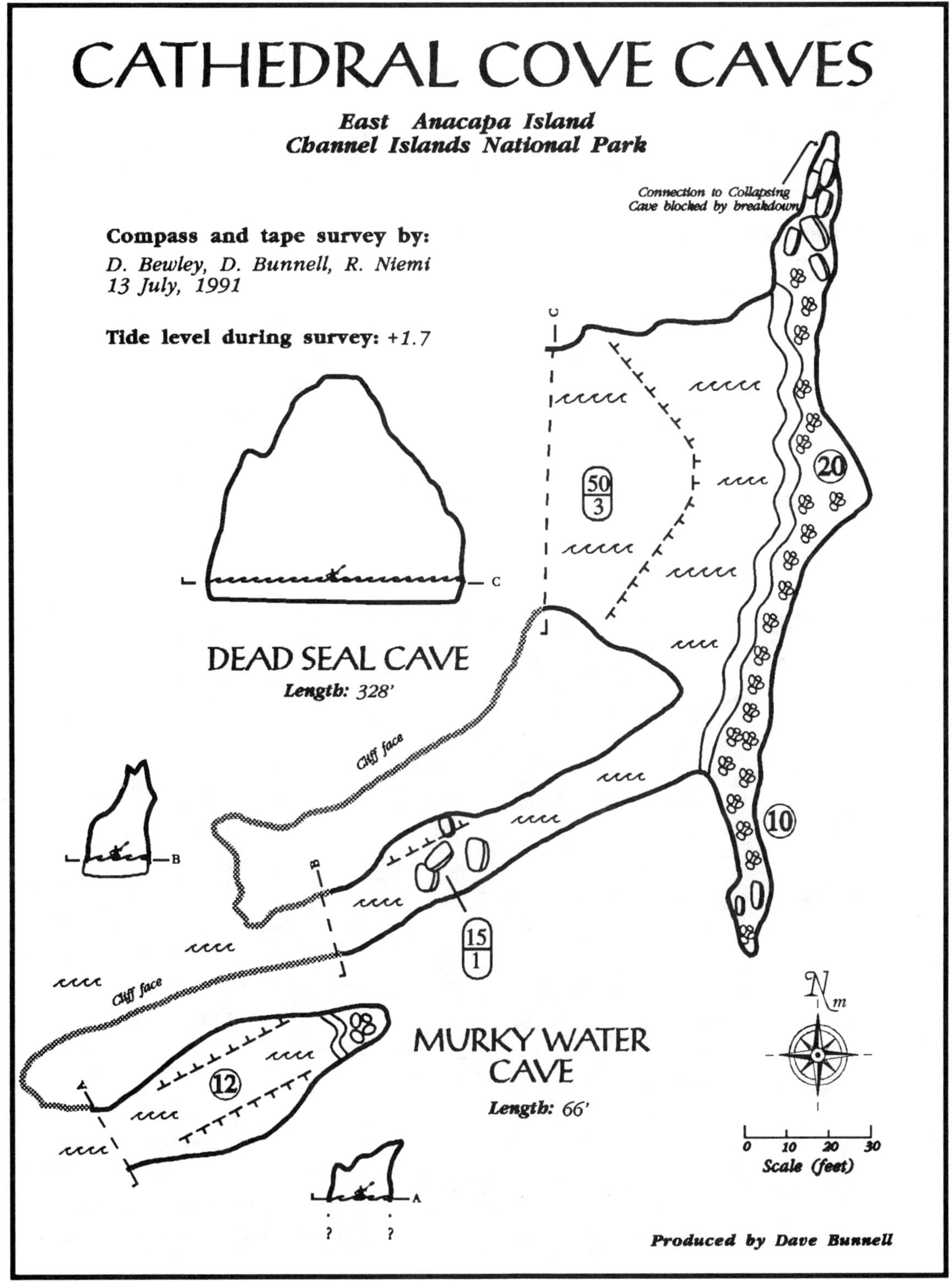
CATHEDRAL COVE CAVES
East Anacapa Island
Channel Islands National Park
Compass and tape survey by:
D. Bewley, D. Bunnell, R. Niemi
13 July, 1991
Tide level during survey: +1.7
Connection to Collapsing Cave blocked by breakdown
DEAD SEAL CAVE
Length: 328'
Cliff face
Cliff face
MURKY WATER CAVE
Length: 66'
0 10 20 30
Scale (feet)
Produced by Dave Bunnell

COLLAPSING CAVE - 13

Location: Cuts through a point on the east end of Cathedral Cove

Entrances: 4

Length: 354'

Conditions: This cave can only be fully explored by swimming and climbing; good footwear and reliable light is necessary. Not suitable for dinghies or kayaks. Unstable rocks pose a hazard.

Description: A very unusual cave with four entrances and a skylight. The cave is formed along four faults. At the convergence of these faults a large breakdown has occurred and formed a skylight. One piece of breakdown is the size of a small house. The main axis runs SW-NE, with two entrances on the east end and one on the west opening into Cathedral Cove. Much of this passage is water floored and narrows to 3' wide. At low tide, the west wall of this passage opens into a fourth entrance. On the east side of the cave's entrance chamber, a squeeze through some breakdown leads into the base of a steep-sided room with walls of dry dirt and broken rock. Climbing up to the right, or south, one crosses over large blocks with pits leading down to the lower level. Several large pieces of driftwood are in this upper level. From the upper level one can climb down a small hole into a lower level which connects back into the entrance chamber.

The western entrance of Collapsing Cave

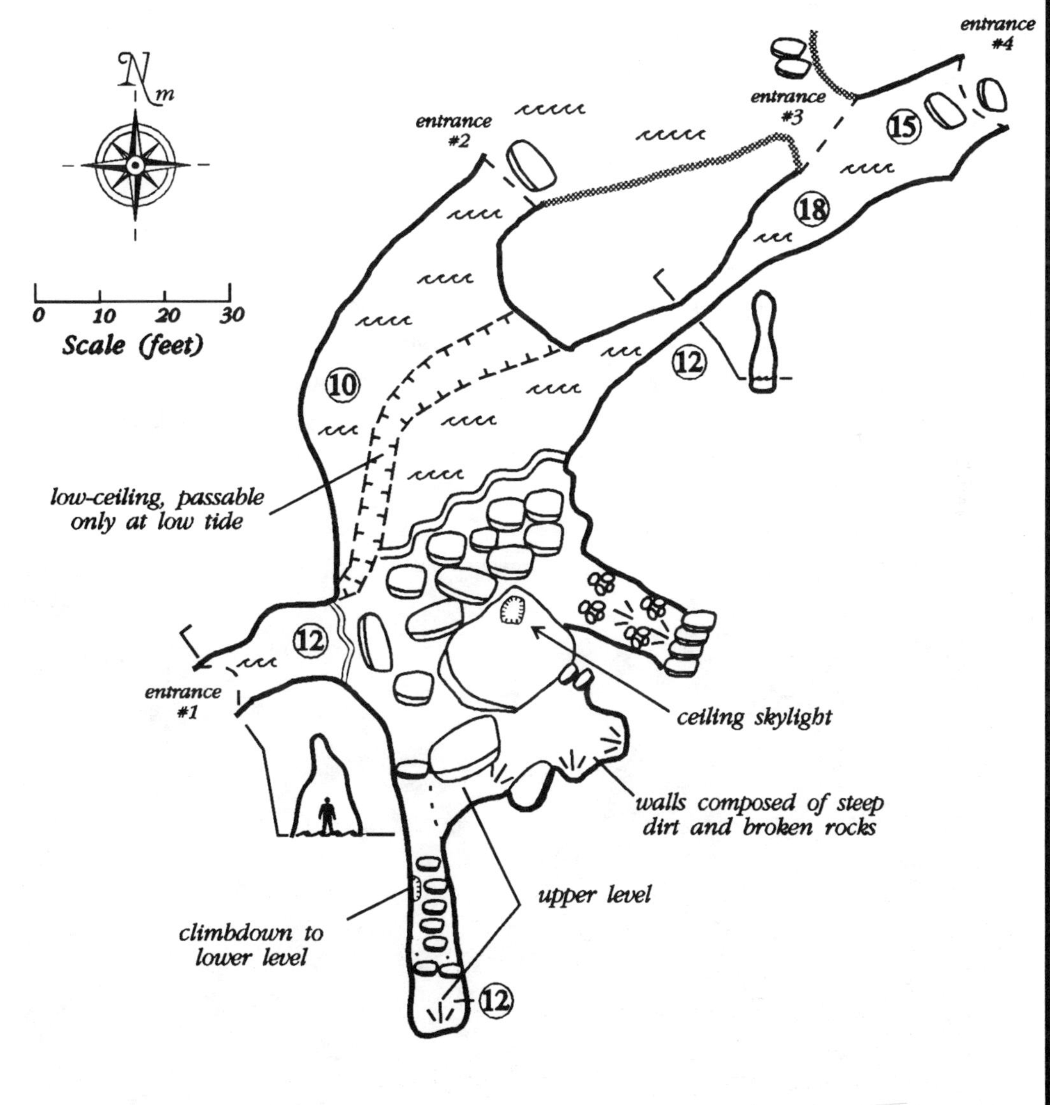
COLLAPSING CAVE
East Anacapa Island, California
Channel Islands National Park
Compass and tape survey by:
D. Bewley, D. Bunnell
7 October, 1990
Cave Length: 354 ft.
Tide level during survey: +0.1
0 10 20 30
Scale (feet)
entrance #1
entrance #2
entrance #3
entrance #4
10
12
12
12
15
18
low-ceiling, passable only at low tide
ceiling skylight
walls composed of steep dirt and broken rocks
upper level
climbdown to lower level
Produced by Dave Bunnell

CLAUSTROPHOBIC BEACH CAVE - 14

Location: This cave and its neighbor are on the west wall of the cove which contains Catacombs Cave, about 1400' west of Landing Cove.

Entrances: 1

Length: 161.6'

Conditions: This cave is entirely water-floored except for a small, low-ceilinged beach in the rear which appears only at low tides. The passage narrows to 6' in places, and there are a few barely submerged rocks, making exploration by dinghy or kayak problematic. Swim or tube in. Lights are useful in the far reaches only.

Description: The entrance is 20' wide and about 10' high. The cave consists of a single passage averaging 6-10' wide and 10-15' high which follows a fault bearing 195°. It ends in a low-ceilinged sandy beach, much of which was in the surf zone at a 1.0 tide level.

TIGHT END CAVE - 15

Entrances: 1

Length: 109.6'

Conditions: This water-floored cave is too narrow for exploration by kayak or dinghy. Only the rear portions require a light and a low tide is useful as well.

Description: The 20' high, rounded entrance to this cave leads into a single passage averaging 12' high and 5' to 6' wide. The fault trend parallels that of neighboring Claustrophobic Beach Cave. At the end the passage pinches into a narrow fissure which extends some 19' before reaching a definitive end.

Claustrophobic Beach Cave entrance

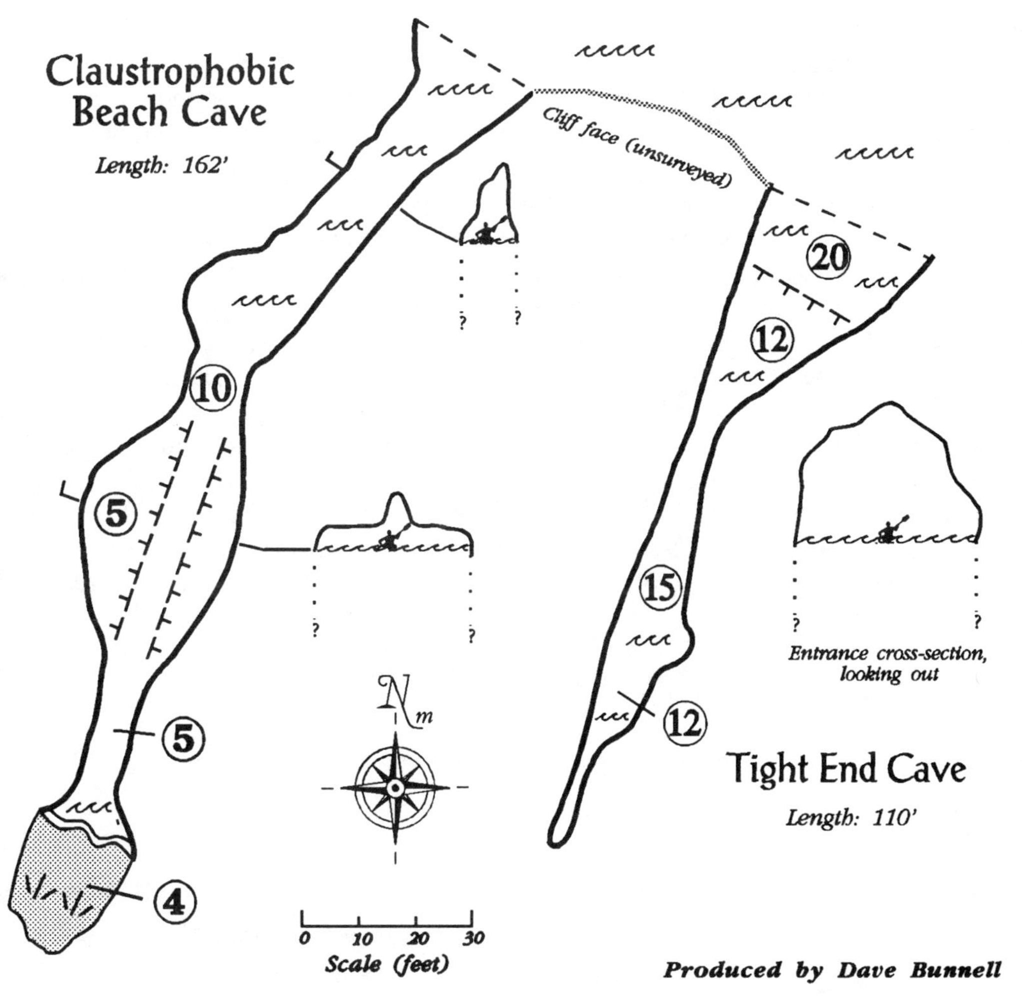
CLAUSTROPHOBIC BEACH
& TIGHT END CAVES
East Anacapa Island
Channel Islands National Park
Compass and tape survey by:
D. Bewley, D. Bunnell, Wm. Liebman
11 June, 1989
Tide level during survey: + 1.0
Claustrophobic Beach Cave
Length: 162'
Cliff face (unsurveyed)
10
5
5
4
20
12
15
12
Entrance cross-section, looking out
Tight End Cave
Length: 110'
N m
0 10 20 30
Scale (feet)
Produced by Dave Bunnell

THE CATACOMBS - 16

Location: 1000' west of Landing Cove

Entrances: 4

Length: 808' (includes 42.6' for cut-off section from 4th entrance.)

Conditions: A good light and a low tide are needed to fully explore this cave, most of which involves walking, crawling, or wading. Much of the cave may even be explorable at high tide, though no doubt requiring some wading. Most of the passages are well protected from surf & swell other than the chamber just inside the right-hand entrance. There is little swimming required except at the westernmost entrance. None of it is suitable for exploration by dinghy or kayak.

Description: This is one of the most interesting and complex caves on the island, connecting three (and at times, 4) entrances and penetrating some 272' into the cliffside. Moving from west to east, entrance #1 is 20' high and extends back through rockfall and surf to a line of fallen boulders. On the left wall is a side passage which reaches a "Y". The left fork leads through two squeezes in a narrow fissure and then to the small middle entrance. The right fork chokes after 30' but appears to have connected to a fourth entrance; the route is now blocked by cobbles and driftwood and may open after storms. Back in the main chamber, the passage becomes low and widens, extending east to interesect another main passage . To the left, the passage extends through a lake, up a 10' climb over fallen rock, and out another large entrance. To the right, the passage extends through another lake to a 20' high chamber floored with small, cemented cobbles. Both lakes contain white sponges. A water-line formed of white deposits and petroleum residue rings this room about 6' off the floor; it continues as well in the passage which extends south. The latter passage is partially blocked by a large rock. Beyond the rock the passage is a narrow, water-floored fissure which lowers to 2 to 3' high for about 30'. At the time of our first visit we were surprised to find two harbor seal pups cowering on the far side of the constriction, and we stopped our exploration. On a later trip we returned and mapped past the low spot, encountering one large harbor seal in a 30' long, 12' high, sandy-floored chamber beyond. The cave owes its complexity to the presence of numerous intersecting faults, and is currently the second longest known cave in the Channel Islands.

CATACOMBS ARCH - 17

Location: Cuts through the east wall of the cove which contains the Catacombs.

Entrances: 2

Length: 58'

Conditions: Swim through or take a kayak if the seas are quite calm; tends to be surgey.

Description: This is a simple tunnel running east-west. It gets up to 25' high and varies from 16 to 6' wide. The cliff wall on the west side shows some interesting erosion patterns,

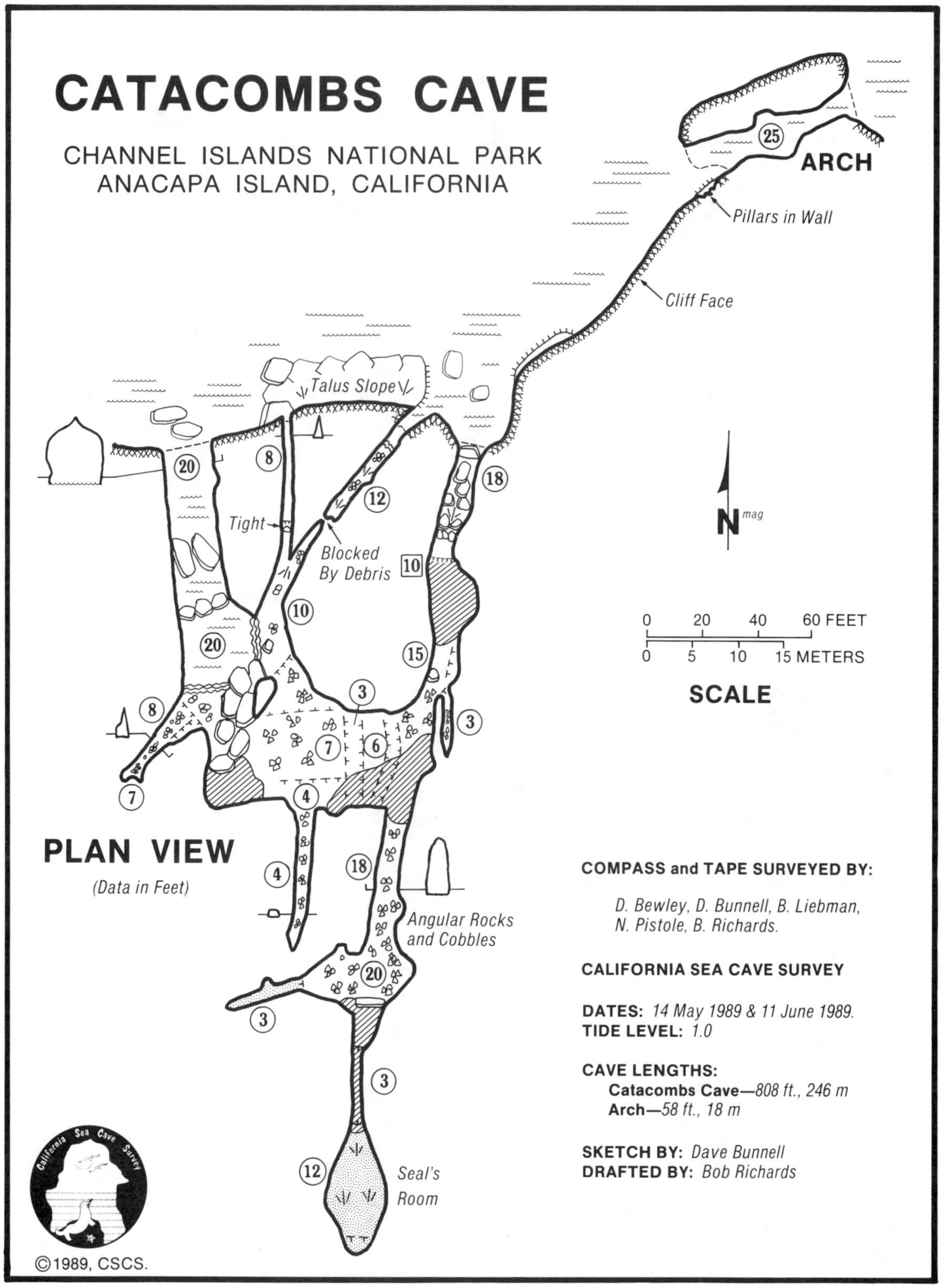
CATACOMBS CAVE
CHANNEL ISLANDS NATIONAL PARK
ANACAPA ISLAND, CALIFORNIA
ARCH
Pillars in Wall
Cliff Face
Talus Slope
Tight
Blocked By Debris
N mag
0 20 40 60 FEET
0 5 10 15 METERS
SCALE
PLAN VIEW
(Data in Feet)
Angular Rocks and Cobbles
Seal's Room
COMPASS and TAPE SURVEYED BY:
D. Bewley, D. Bunnell, B. Liebman, N. Pistole, B. Richards.
CALIFORNIA SEA CAVE SURVEY
DATES: 14 May 1989 & 11 June 1989.
TIDE LEVEL: 1.0
CAVE LENGTHS:
Catacombs Cave—808 ft., 246 m
Arch—58 ft., 18 m
SKETCH BY: Dave Bunnell
DRAFTED BY: Bob Richards
California Sea Cave Survey
©1989, CSCS.

Section of cliff showing the entrances to Catacomb Cave. The main or dry entrance is visible on the far left, with its two prominent faults diverging towards the top. The crawlway entrance is barely visible, and the wet entrance is seen about a third of the way in from the right edge of the photo.

In the dark interior of the Catacombs. The rock conceals the passageway that leads into the Seal's Room.

Climb-down below the main entrance to the Catacombs

The impressive main passage of Catacombs Cave

DEBRIS CAVE - 18

Location: At the rear of a small cove 900' west of Landing Cove.

Length: 232'

Entrances: 1

Conditions: The cave is floored with water throughout but is generally too shallow for exploration by dinghy or kayak. Swim or tube in at low tide and bring a light.

Description: The cave has two entrances. The lefthand entrance is low and only useable at low tide. The larger right-hand entrance is 20' high and marked by a large boulder, covered with red algae. The passages lead into a single passage averaging 16 to 20' wide and 10 to 20' high which extends to a 20' long sloping cobble beach. On the rear wall a 12' long fissure continues along the main fault trend. Here particularly, and throughout the cave are found various flotsam, including portions of diving fins and knife, and numerous pieces of hose wedged in the rocks.

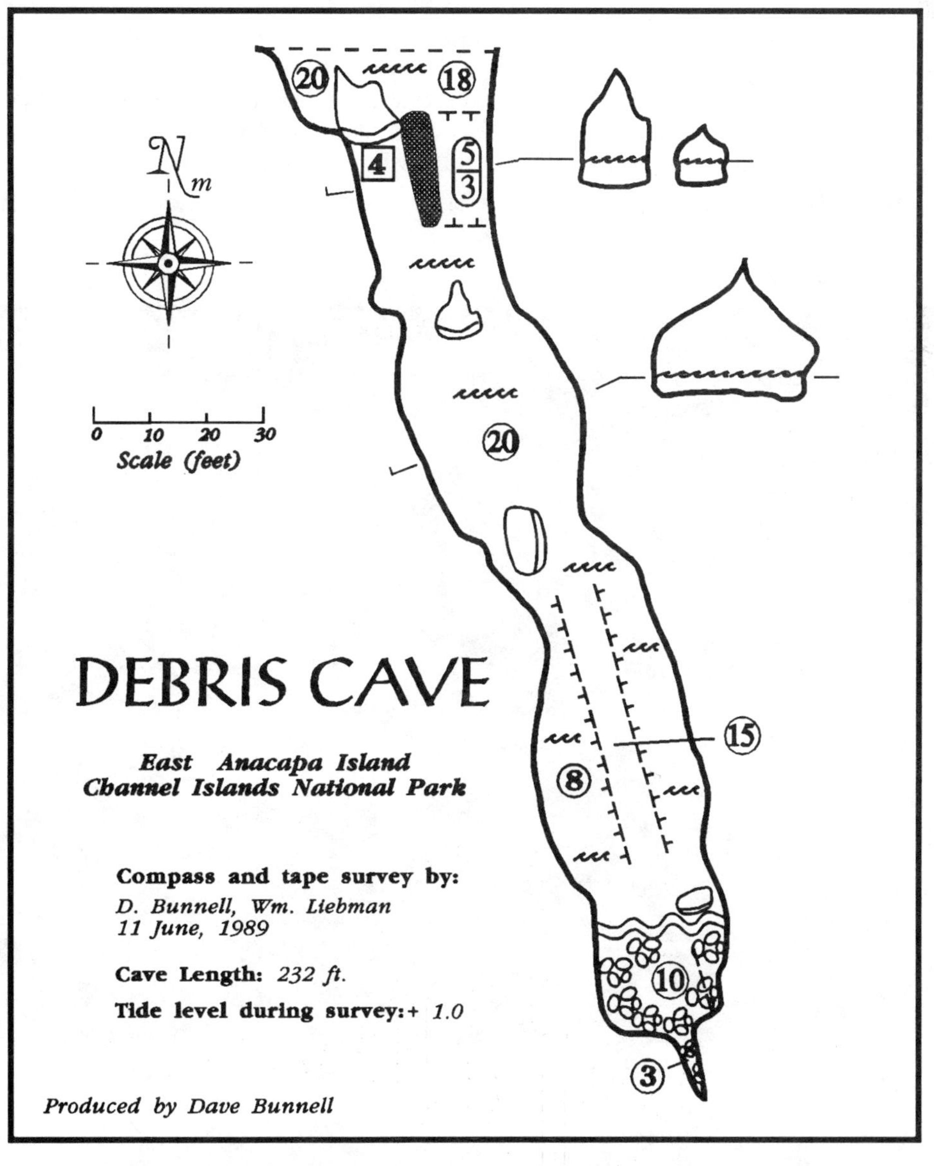

HALF-GONE CAVE - 19

Location: Entrance #1 is about 200' west of Landing Cove. The western entrance in the collapsed portion is about 500' west of Landing Cove.

Entrances: 3

Length: 462' (141 m).

Conditions: At a moderate or high tide, one can enter the cave by kayak through the middle entrance (just east of Garbage Cove) and paddle into the main chamber, which is cut off by rockfall from the open sea. This makes it a very sheltered trip for a kayak, and a safe place to turn around. At low tide, submerged rocks are uncovered and kayaking is problematic.

Description: There are three entrances, two of which are visible from the sea. The eastmost is partially obscured by rockfall but opens into an impressive chamber some 40' high and 60' long, divided by a large pillar. A 60' swim leads into a second water-floored chamber with a large entrance. Continuing along the same SW trend one emerges at the base of a large collapse. Climbing this leads one into a large littoral sinkhole, or collapsed portion of the cave. That this was once part of the cave is suggested by the presence of an isolated portion of cave on the SW corner of the sinkhole, which continues on the same fault axis as the main cave. This sinkhole is known to Park Service personnel as Garbage Cove, since it was once used by the Coast Guard as a rubbish dump in days past Apparently a 'dozer operator drove off the top into here and was killed. Remains of the dozer are still visible.

Kayakers in the inner chamber of Half-Gone at high tide

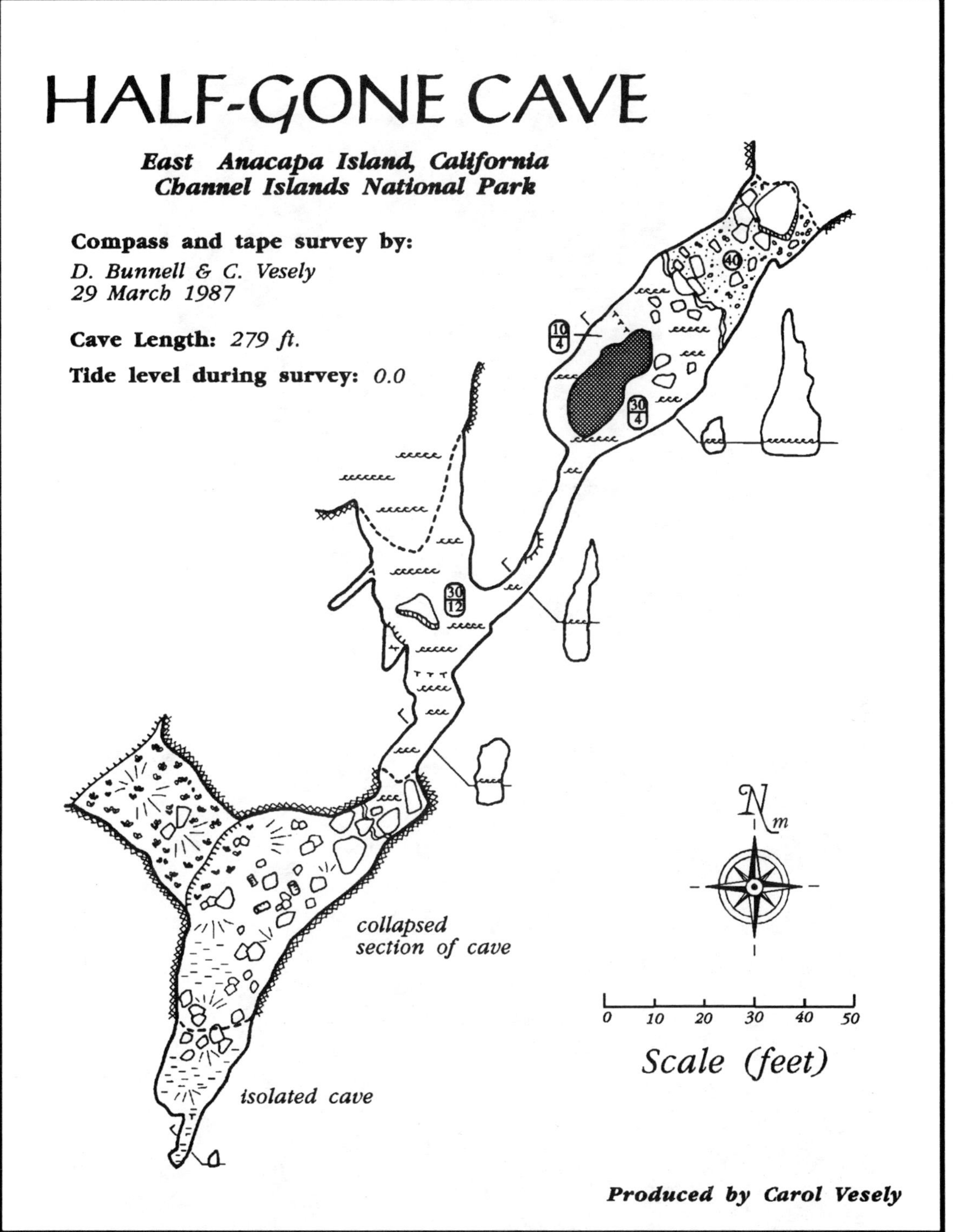
HALF-GONE CAVE
East Anacapa Island, California
Channel Islands National Park
Compass and tape survey by:
D. Bunnell & C. Vesely
29 March 1987
Cave Length: 279 ft.
Tide level during survey: 0.0
collapsed
section of cave
isolated cave
0 10 20 30 40 50
Scale (feet)
Produced by Carol Vesely

Half Gone Cave, eastern or rockfall entrance

Inside the main chamber of Half-Gone Cave

Djuna Bewley "tubing" in Rockfall Cave (next page)

ROCKFALL CAVE - 20

Location: These caves are about 150' west of Landing Cove; this one is mostly obscured by rockfall

Entrances: 1 **Length**: 105'

Conditions: Exploration involves climbing through breakdown and swimming or wading deep pools in darkness. Bring a good light. The pool height varies with the tide level.

Description: The outer portions of this cave have collapsed, leaving a large breakdown pile. One enters by doing a 10' climbdown on the left side (facing in) or alternatively, by squeezing through the rocks along the left wall. The cave consists of a linear single passage averaging 10' wide and 15' tall. It is floored with water for most of its length.

OUTSIDE TUNNEL CAVE - 21

Entrances: 2 **Length:** 353'

Conditions: This cave is entirely water-floored. Portions might be explored by dinghy but the passage narrows to 6 to 7' wide in spots and there are some submerged rocks. Swimming or tubing is probably the most optimal means of exploration. Bring a light and explore at low tide.

Description: The cave has two entrances. The left-hand entrance is narrow and easily goes unnoticed from offshore. It leads into a 2-3' wide fissure which extends some 30' before opening into a larger chamber. The more prominent right-hand entrance, some 40' high, leads into the same chamber. On the right side of this chamber are several faults and fissures, one of which leads into a low passage which connects back into the cave's main passage, and follows the same fault as the narrower outside tunnel. It averages 10 to 15' high and 6 to 7' wide and leads to a small cobble beach with a narrow 20' long fissure extending the cave further into the cliff. There are numerous submerged rocks.

The two entrances to Outside Tunnel. Rockfall Cave is behind fallen rocks on the right.

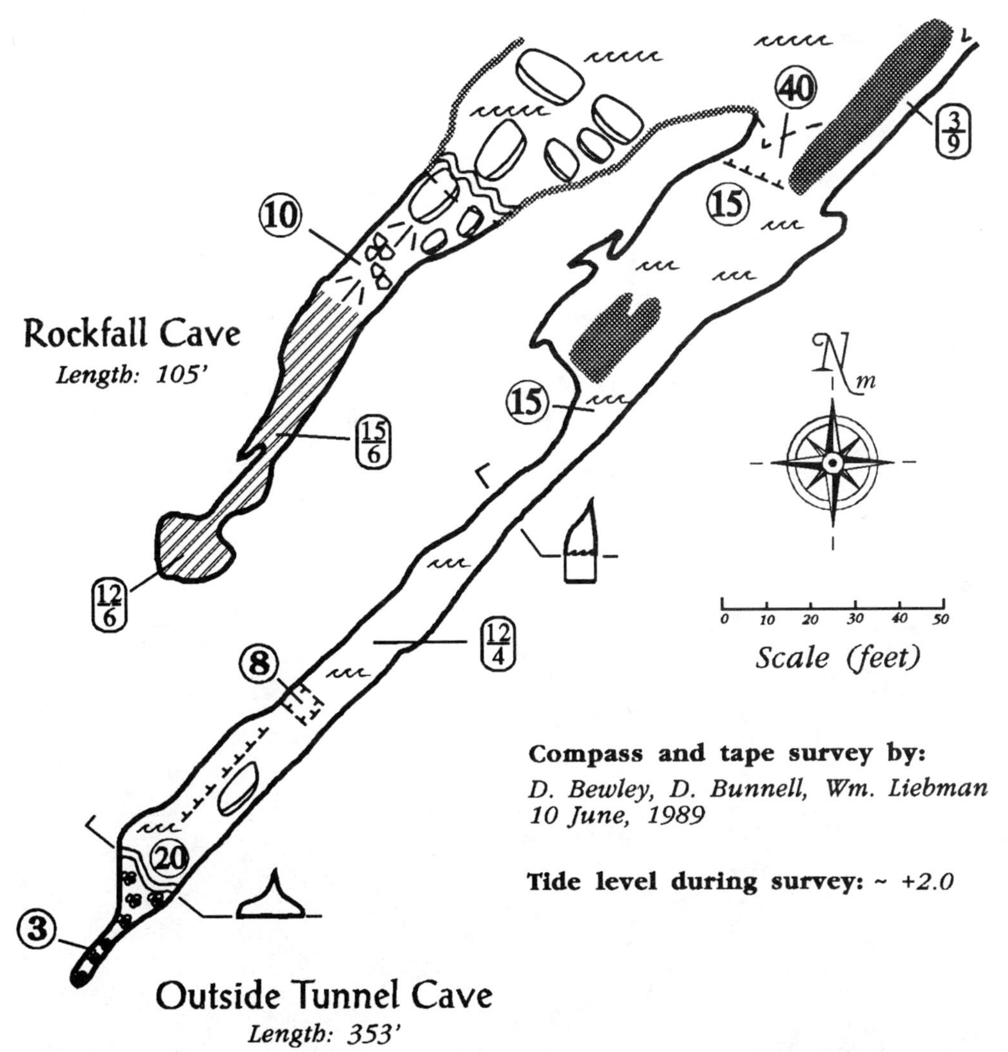
ROCKFALL & OUTSIDE TUNNEL CAVES
East Anacapa Island, California
Channel Islands National Park
Rockfall Cave
Length: 105'
Outside Tunnel Cave
Length: 353'
10
15
40
15
8
20
3
Scale (feet)
0 10 20 30 40 50
Compass and tape survey by:
D. Bewley, D. Bunnell, Wm. Liebman
10 June, 1989
Tide level during survey: ~ +2.0
Produced by Dave Bunnell

RECTANGULAR REFLECTIONS CAVE - 22

Location: About 120' west of Landing Cove

Entrances: 1

Length: 154'

Conditions: Explore at low tide by swimming or tubing in. Bring lights.

Description: The entrance is 30' high but soon closes to a rectangular pasage some 2' wide and 7' high above the water. Light streaming through casts rectangular relections on the water within. The passage widens after 15' and continues as a single water-floored passage which varies from 5' to 8' high. It is about 8' wide for most of its length with no beach in the rear.

PENDANT CAVE - 23

Location: About 110' west of Landing Cove

Entrances: 1

Length: 44'

Conditions: Water-floored, no lights needed. One may enter with a small kayak or dinghy at a low tide if sea conditions are favorable.

Description: The entrance is 20' high and 20' wide. The passage is split by a long ceiling pendant which probably forms two separate passages at high tide. In the back lefthand corner of the cave a fissure passage continues and the presence of air movement here suggests that it may continue for some distance, but is too tight to follow on the water's surface.

Pendant Cave on the left; Rectangular Reflections on the right (just to left of kayaker's head)

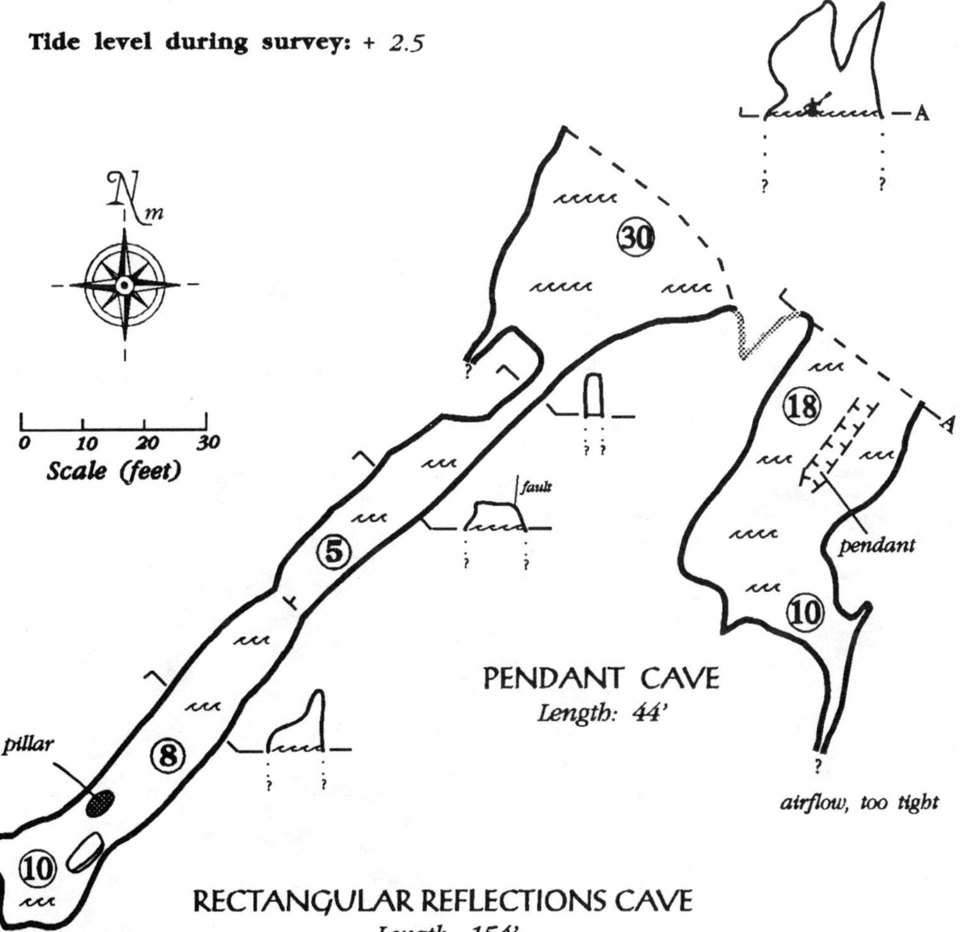
RECTANGULAR REFLECTIONS
& PENDANT CAVES
East Anacapa Island, California
Channel Islands National Park
Compass and tape survey by:
D. Bewley, D. Bunnell, Wm. Liebman
10 June, 1989
Tide level during survey: + 2.5
0 10 20 30
Scale (feet)
pillar
fault
pendant
airflow, too tight
PENDANT CAVE
Length: 44'
RECTANGULAR REFLECTIONS CAVE
Length: 154'
Produced by Dave Bunnell

LANDING COVE CAVE - 24

Location: In the rear (south) wall of Landing Cove. Indicated on the USGS topographic map.

Entrances: 1

Length: 148'

Conditions: Water-floored but fairly spacious. A dinghy or kayak could be taken just inside, probably even at high tide. Tends to be pretty surgey. A light would be needed only in the rear portions.

Description: The cave is a single water-floored passage formed along two parallel faults. Ceiling heights range from 12 to 25'. Also in Landing cove is a short cave behind the landing dock itself, which forms a blowhole much of the time and is unenterable. It appears to have two entrances.

Landing Cove Cave lies on a very prominent fault

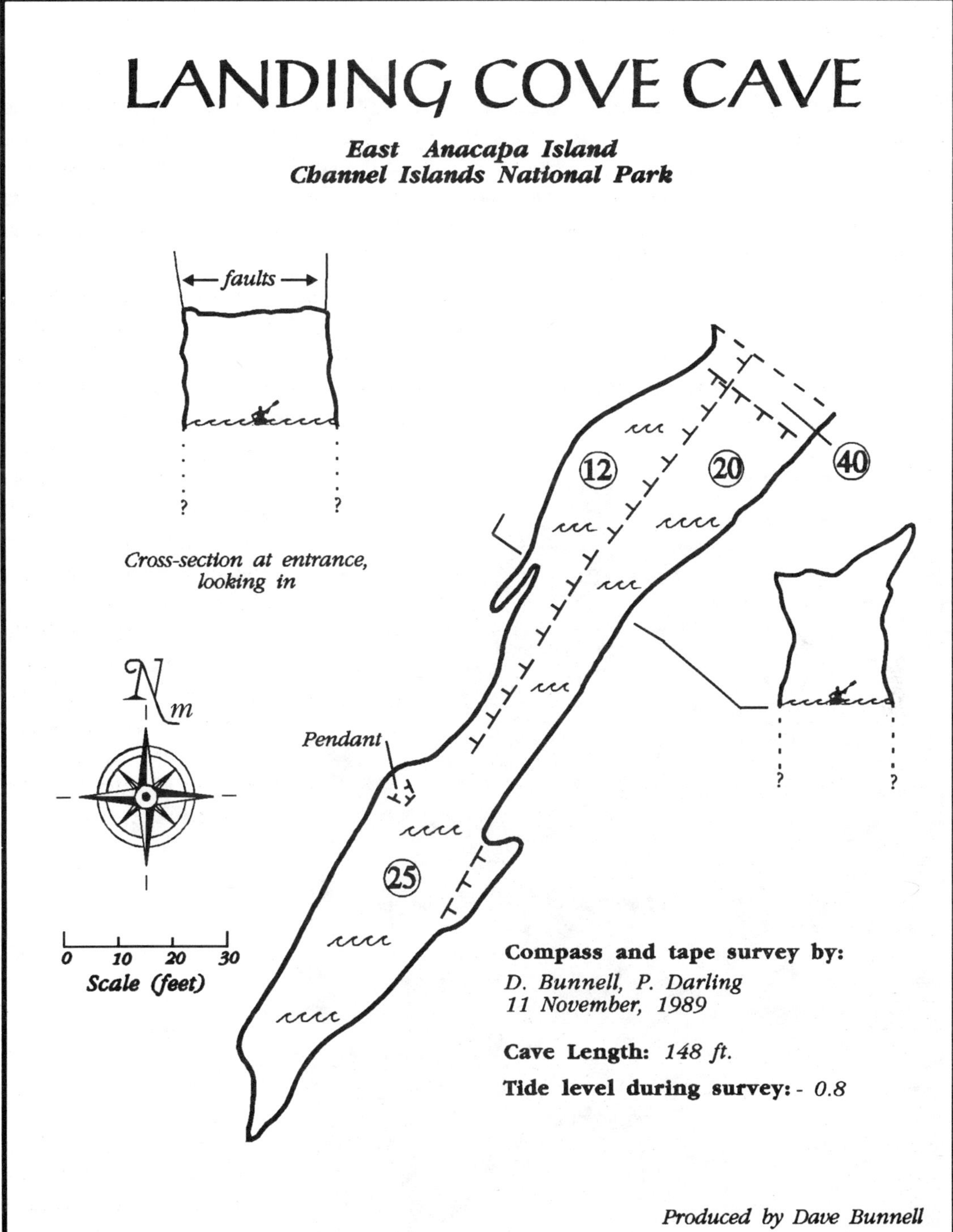
LANDING COVE CAVE
East Anacapa Island
Channel Islands National Park
faults
?
?
Cross-section at entrance, looking in
12
20
40
Pendant
25
?
?
N
m
0 10 20 30
Scale (feet)
Compass and tape survey by:
D. Bunnell, P. Darling
11 November, 1989
Cave Length: 148 ft.
Tide level during survey: - 0.8
Produced by Dave Bunnell

LA GRIETA (The Fissure) - 25

Location: About 250' west of Arch Rock

Entrances: 1

Length: 381'

Conditions: A low tide is probably not so important here as calm seas. Some of the passage in the entrance section is quite narrow, and at low tide sharp barnacles are uncovered. A higher tide entry would keep one above the barnacles, but with the lowering in ceiling height, calm seas would be needed. Beyond the constriction the swells are diminished. An entry with SCUBA is possible if one has training in cavern diving (the first 40' are roofed over and hence constitute a cave dive).

Description: This unusual cave is formed along a fault running somewhat parallel to the cliff face. Above water, the entrance appears as a 30' high narrow fissure; below water, the fissure widens and extends to 20' deep depending on the tide level. Entry into the above-water portions is through a 25' long section of narrows with ledges on which one can keep above water. Beyond the passage widens to nine feet and continues as a water-floored passage some 300' to a cobble beach. Beyond the beach the passage forks into two smaller tunnels which rejoin in a terminal room, which contained little other than some small dripstone formations and a seal skull. This cave has the deepest straight-line penetration into the cliff of any cave on Anacapa.

La Grieta's entrance is a tall fissure. Note the small waterfall pouring off of the narrow shelves just inside the entrance.

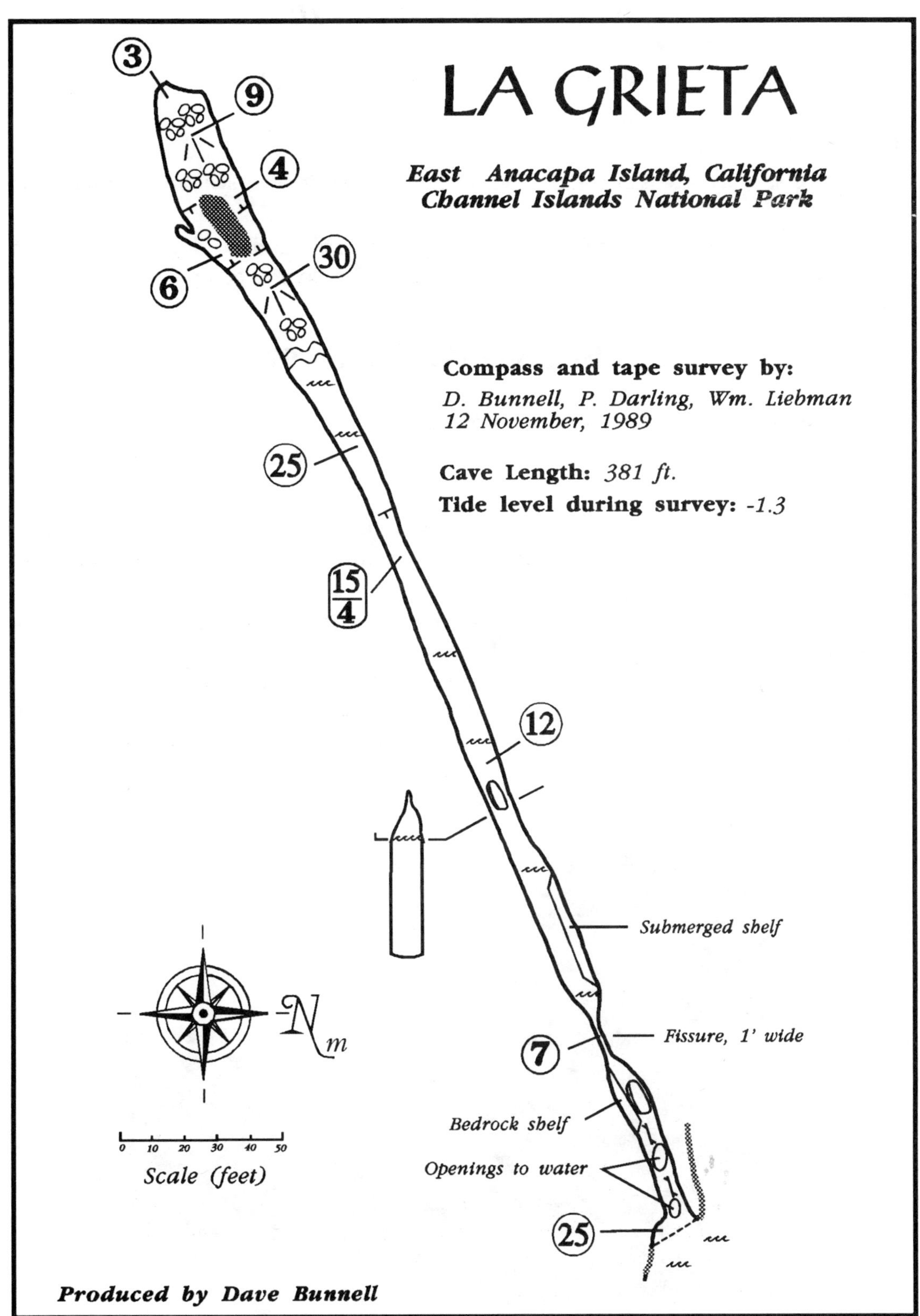
LA GRIETA
East Anacapa Island, California
Channel Islands National Park
Compass and tape survey by:
D. Bunnell, P. Darling, Wm. Liebman
12 November, 1989
Cave Length: 381 ft.
Tide level during survey: -1.3
3
9
4
30
6
25
15
4
12
Submerged shelf
Fissure, 1' wide
7
Bedrock shelf
Openings to water
25
N
m
0 10 20 30 40 50
Scale (feet)
Produced by Dave Bunnell

EAST END CAVE - 26

Location: On the north shore near the eastern tip of East Island

Entrances: 1

Length: 46'

Conditions: This cave is dry at low tide, so one can walk in after swimming or kayaking to it.

Description: Despite having a large entrance (25' wide and 30' high), the passage narrows to 5' wide some 20' in from the dripline. The floor is composed of unusually large cobbles.

EAST END TUNNEL - 27

Location: Cuts through the eastern tip of East Island

Entrances: 3

Length: 128'

Conditions: The cave is largely dry at low tide, and if the seas are calm a dinghy might be landed on the south side. At higher tide a dinghy might conceivably pass through. No lights are needed.

Description: This is a scenic little cave in a spectacular location. The north shore entrance affords an outstanding view of the sea stacks around the arch. The cave has three entrances. The main passage is some 12' high and cobble floored, cutting through the island on a NW-SE axis. The passage varies from 26' wide at the south entrance to 16' wide on the north. The third entrance is accessed from an east-trending passage which is a fissure, 12' high and 6' wide.

The south shore entrance to East End Tunnel

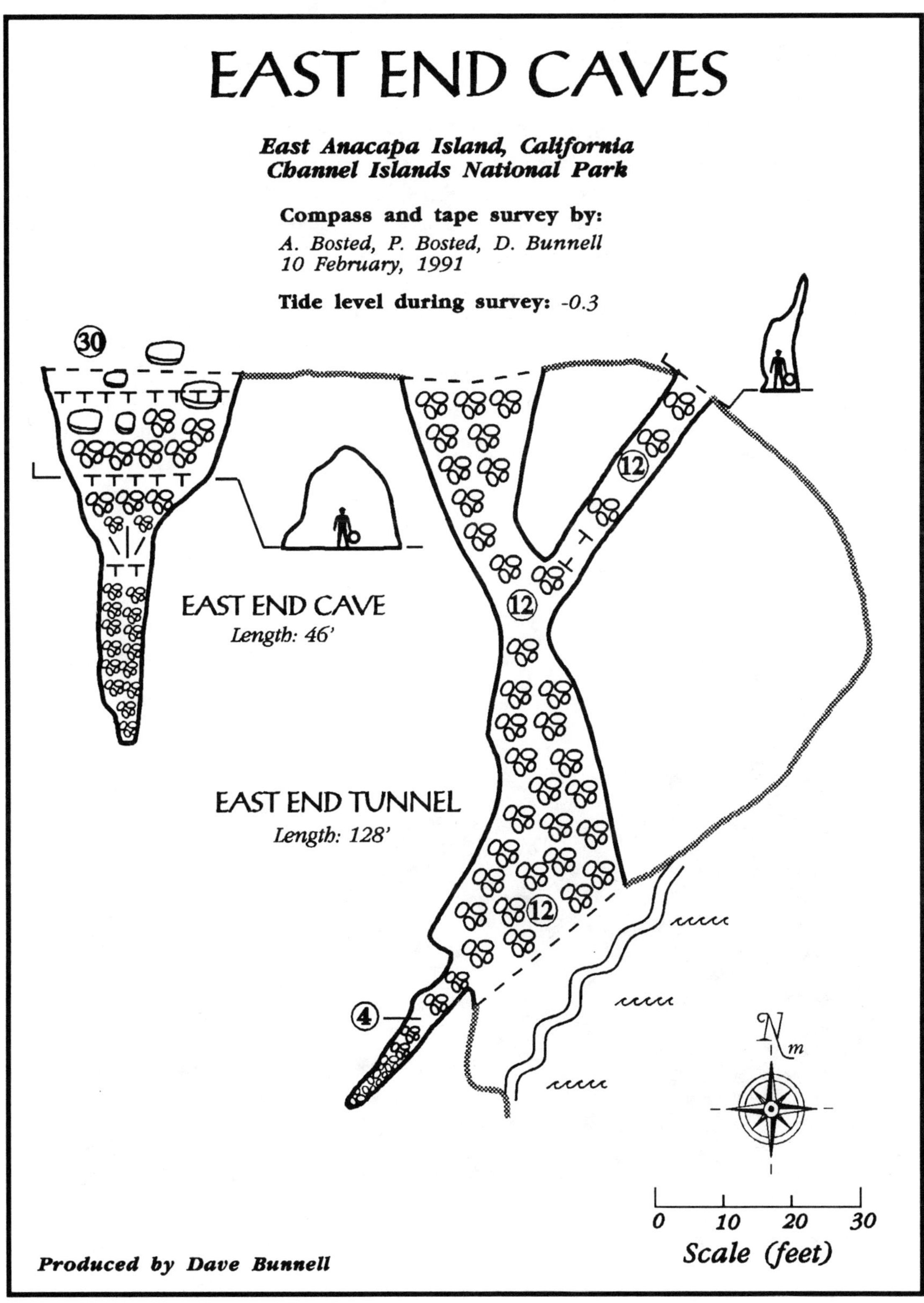

EAST END CAVES
East Anacapa Island, California
Channel Islands National Park
Compass and tape survey by:
A. Bosted, P. Bosted, D. Bunnell
10 February, 1991
Tide level during survey: -0.3
30
12
12
12
4
EAST END CAVE
Length: 46'
EAST END TUNNEL
Length: 128'
N m
0 10 20 30
Scale (feet)
Produced by Dave Bunnell

ARCH ROCK CAVES - 28 & 29

Location: In the two limbs of the large arch at the east end of East Island

Entrances: 2 (for each cave)

Lengths: Western Cave (75'); eastern cave (48')

Conditions: Both caves afford "through-trips", which in the larger western cave can be done in a kayak or dinghy if the seas are calm. They also make an interesting snorkel, conditions permitting. At low tide submerged rocks at the NW entrance to the west cave produce an intermittent waterfall which could pose problems for a dinghy. No lights are needed.

Description: The great arch at the end of east island is one of the most prominent features of the Anacapas. Besides the great arch itself, there are tunnels cutting through each limb of the arch. The cave on the western limb is the larger and is "L"-shaped. The chamber inside is some 25' wide and up to 20' high (at low tide). At low tide it becomes quite shallow and it is possible to wade in portions of the room. The easternmost cave is in deeper water and consists of a 20' high fissure varying from 8-12' wide, gently curving.

Arch Rock, seen from the south side. Cave entrances are visible in both limbs of the arch, and one can see through the tunnel in the right limb.

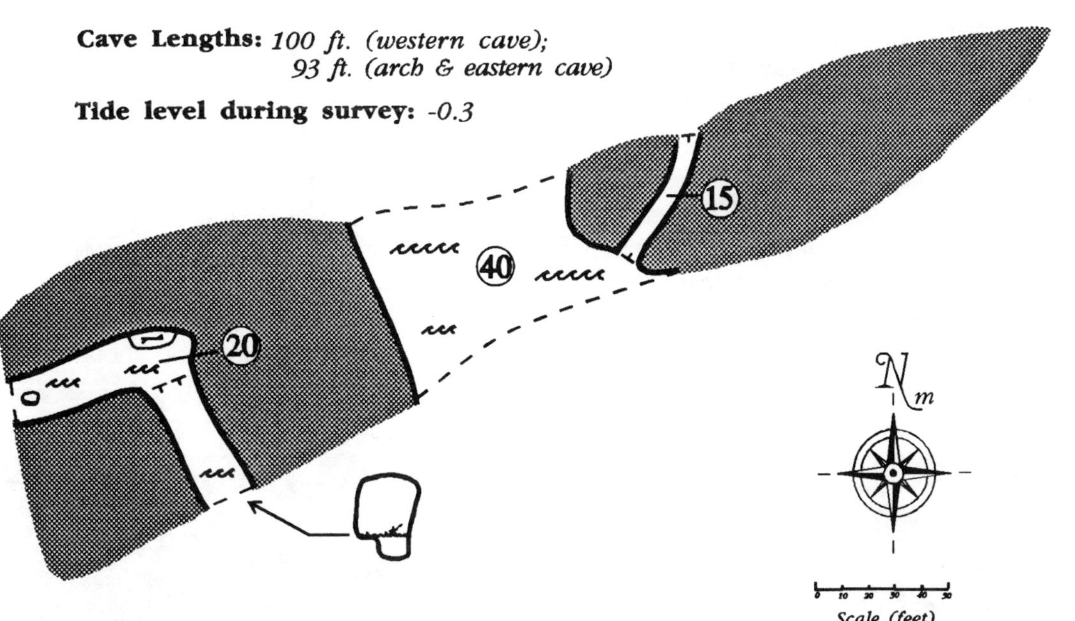
CAVES OF ARCH ROCK
East Anacapa Island, California
Channel Islands National Park
Compass and tape survey by:
P. Bosted, D. Bunnell
10 February, 1991
Cave Lengths: 100 ft. (western cave);
93 ft. (arch & eastern cave)
Tide level during survey: -0.3
15
40
20
Nm
Scale (feet)
Produced by Dave Bunnell

FEARFUL FISSURE - 30

Location: On the south shore of East Anacapa, below the lighthouse.

Entrances: 1

Length: 157

Conditions: Calm seas and lights are needed to visit this cave.

Description: A large rock partially blocks the entrance. The cave is a single passage, averaging some 12' high, largely water-floored, ending in a cobble beach in the rear. This was the only cave described here that was not personally entered by the author.

LIGHTHOUSE CAVE - 31

Location: On the south shore of East Anacapa, below the lighthouse

Entrances: 2

Length: 218

Conditions: A moderate tide level and calm seas are required. A light is useful in the rear of the main passage. A through-trip in a kayak or dinghy is possible with a low tide and calm seas.

Description: Two entrances feed into a large passage, 15' high and 10' wide, which extends to a sand beach. The eastern entrance is the smaller of the two, and may be largely be submergered at high tide. The larger western entrance is somewhat more sheltered from swell. Just east of this cave is a shallow cave with a large entrance. The day we surveyed it, surf was bouncing violently off the back wall and it was named "Rebound Cave". Since we did not "push" it, it may extend further than shown on the map.

Caves below the lighthouse. Lighthouse Cave is the small entrance on the left, Rebound Cave is in the center, and the Fearful Fissure is on the right, behind a large rock.

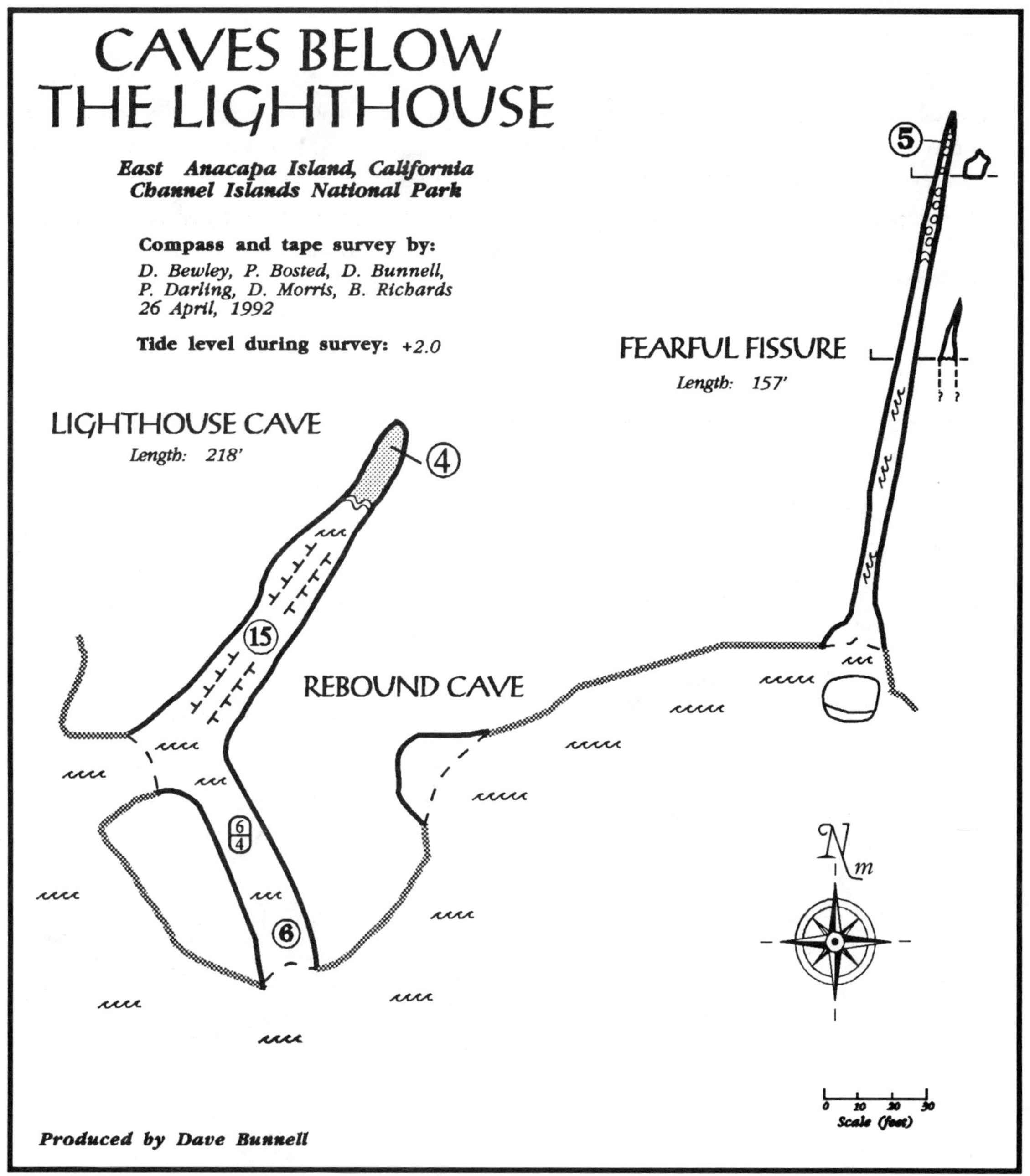
CAVES BELOW
THE LIGHTHOUSE
East Anacapa Island, California
Channel Islands National Park
Compass and tape survey by:
D. Bewley, P. Bosted, D. Bunnell,
P. Darling, D. Morris, B. Richards
26 April, 1992
Tide level during survey: +2.0
FEARFUL FISSURE
Length: 157'
LIGHTHOUSE CAVE
Length: 218'
REBOUND CAVE
Scale (feet)
Produced by Dave Bunnell

LOST GLASSES CAVE - 32

Location: 2000' east of the western tip of East Island.

Entrances: 1

Length: 72'

Conditions: The entire cave is dry at low tide and is probably enterable at any tide level. A dinghy or kayak may be landed directly into the cave or at the beach just to the west. However, if the surf is rough, hang on to your glasses!

Description: The cave is one large sandy-floored room, with a large entrance 48' wide and 25' high. The passage necks down to a small fissure in the rear which is choked with cobbles, buoys, and other detritis.

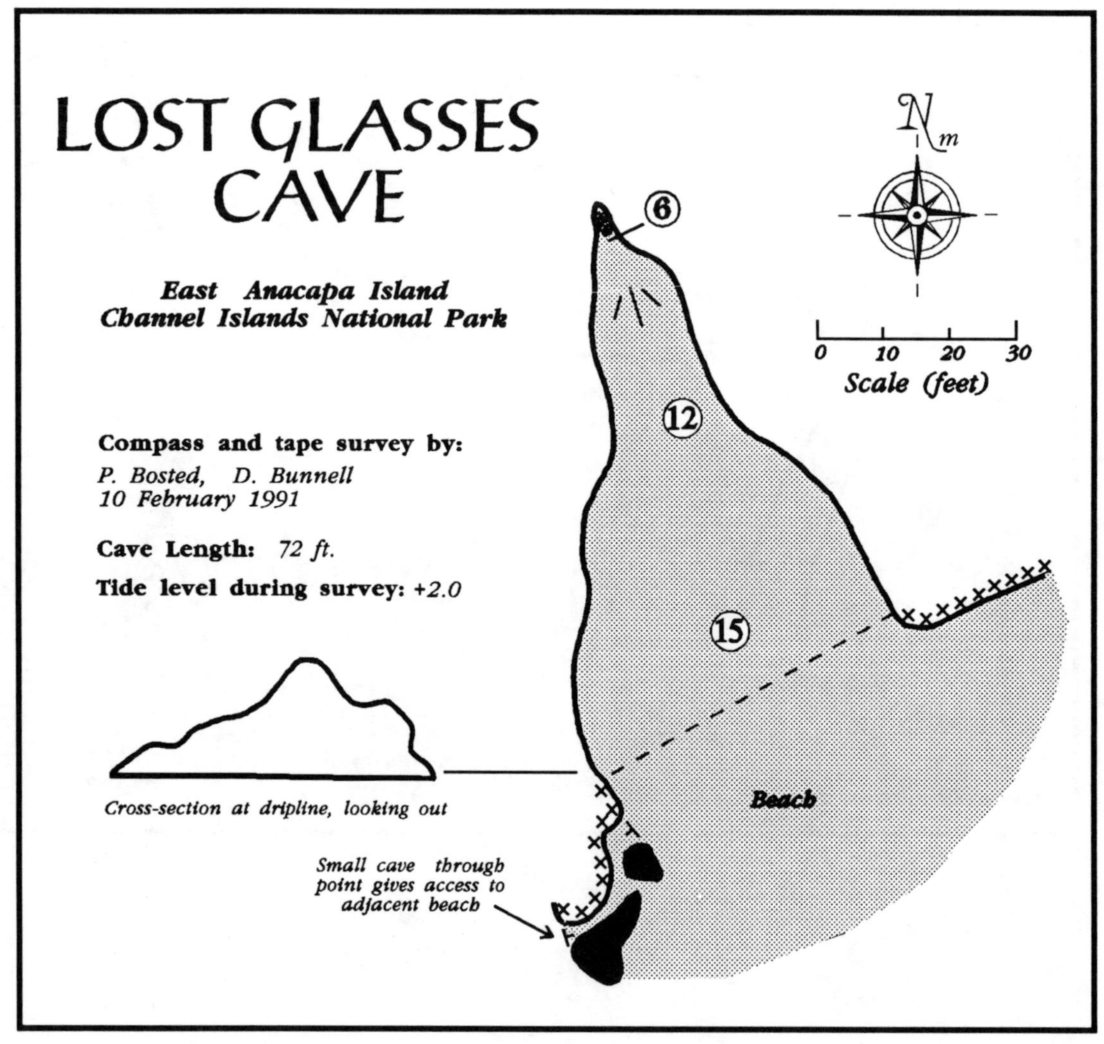

TIP CAVE - 33

Location: 200' east of the western tip of East Island.

Entrances: 1

Length: 41'

Conditions: The entire cave is dry at low tide. A dinghy or kayak may be landed at the beach just to the west and one can then walk into the cave.

Description: The cave is a single chamber floored with sand and large cobbles. There is a nice view from the entrance.

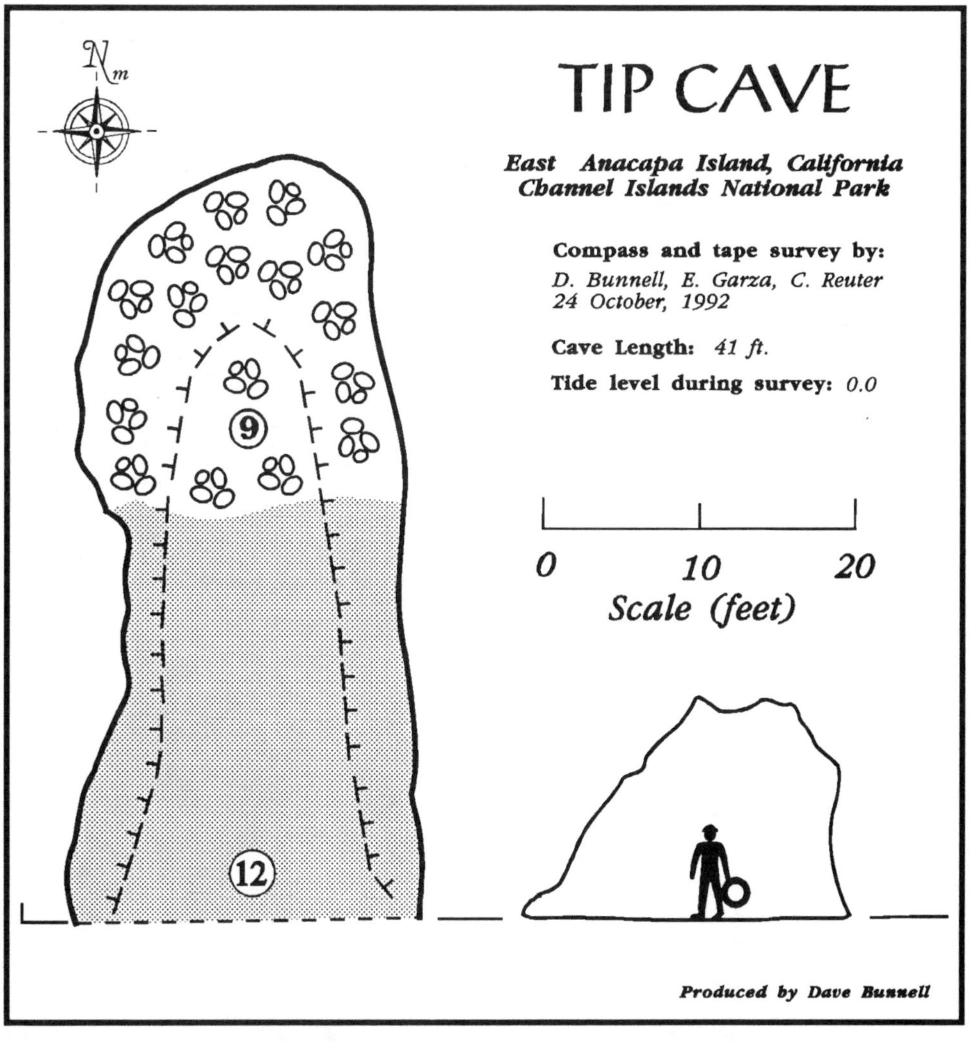

TIP CAVE

East Anacapa Island, California
Channel Islands National Park

Compass and tape survey by:
D. Bunnell, E. Garza, C. Reuter
24 October, 1992

Cave Length: *41 ft.*

Tide level during survey: *0.0*

0 10 20
Scale (feet)

N m

9

12

Produced by Dave Bunnell